Subsurface Contamination Monitoring Using Laser Fluorescence

Subsurface Contamination Monitoring Using Laser Fluorescence

Edited by

Katherine Balshaw-Biddle

Rice University, Houston, TX

Carroll L. Oubre

Rice University, Houston, TX

C. Herb Ward

Rice University, Houston, TX

Authors

Jonathan E. Kenny

Jane W. Pepper

Andrew O. Wright

Yu-Min Chen

Steven L. Schwartz

Charles G. Shelton

Acquiring Editor: Skip DeWall
Project Editor: Ibrey Woodall
Cover design: Jonathan Pennell

Library of Congress Cataloging-in-Publication Data

Subsurface contamination monitoring using laser fluorescence /
edited by Katherine Balshaw-Biddle, Carroll L. Oubre, C.H. Ward.
p. cm. — (AATDF monograph series)
Includes bibliographical references and index.
ISBN 1-56670-481-2 (alk. paper)
1. Soil—Pollution—Measurement. 2. Groundwater flow—
Pollution—Measurement. 3. Hazardous waste site remediation—Technological innovations.
I. Balshaw-Biddle, Katherine. II. Oubre, Carroll L. III. Ward, C.H. (Calvin Herbert), 1933-. IV. Series.
TD878.S833 1999
628.5′028′7—dc21

99-26361
CIP

Although the information described herein has been funded wholly or in part by the United States Department of Defense (DOD) under Grant No. DACA39-93-1-0001 to Rice University for the Advanced Applied Technology Demonstration Facility for Environmental Technology Program (AATDF), it may not necessarily reflect the views of the DOD or Rice University, and no official endorsement should be inferred.

International Standard Book Number 1-56670-481-2
Library of Congress Card Number 99-26361
Printed in the United States of America 1 2 3 4 5 6 7 8 9 0
Printed on acid-free paper

Foreword

The U.S. Department of Defense/Advanced Applied Technology Development Facility (DOD/AATDF) selected the Tufts laser-induced fluorescence (LIF) project, *Subsurface Contaminant Monitoring by Laser Excitation-Emission Matrix and Cone Penetrometer,* as the one monitoring study in its 12 projects. It built upon the past EPA-funded LIF research of the Principal Investigator, Dr. Jonathan E. Kenny, and complemented the DOD research interests of the Tri-Service Characterization and Analysis Penetrometer System (SCAPS) program, coordinated by Mr. George Robitaille of the Army Environmental Center. Dr. Kenny was able to leverage the AATDF funding by securing discounted instrumentation prices and donated equipment from selected suppliers, cost sharing from Tufts University, and a publishing contract for the fluorescence database to be generated during the project.

Several organizations and companies that possessed cone penetrometer testing (CPT) capabilities were approached about participation in the project. The selection of the CPT provider, Fugro Geosciences, Inc., and Fugro Environmental, was based upon low proposed costs for the field tests, working knowledge of commercial LIF systems, and potential interest in licensing the technology. Fugro Geosciences also provided assistance in system design, fabrication, and integration of the probe with the CPT electronics and equipment. Fugro Environmental provided assistance with preparation of the Work Plan Addendum and the Technology Evaluation Report in addition to management of field support.

Two SCAPS researchers, Dr. Ernesto Cespedes, U.S. Army Engineer Waterways Experiment Station (USAE WES), and Dr. Steve Lieberman, Naval Research and Development (NRAD), agreed to be shepherds at the inception of this project to ensure that the Tufts research complemented and did not duplicate the SCAPS LIF research.

The *Subsurface Contaminant Monitoring by Laser Excitation-Emission Matrix* project was led by AATDF Project Manager, Dr. Kathy Balshaw-Biddle. The research was conceived and performed by the team of Professor Jonathan E. Kenny (Tufts University), which included Ms. Jane Pepper, Dr. Andrew Wright, Dr. Yu-Min Chen, Dr. Sean Hart, Dr. Sam Mathew, Dr. Ranjith Premasiri, and Mr. Todd Pagano. The engineering consultant and commercialization advisor on this project was Mr. Steven Schwartz, Fugro Environmental. Mr. Recep Yilmaz, president of Fugro Geosciences, and Mr. Robert Benning, Regional Manager for Fugro Environmental, provided company resources to support the project. Assistance with probe fabrication and field implementation was provided by personnel at Fugro Geosciences, Inc., including Mr. Dennis Stauffer and Mr. Andrew Taer.

Project shepherds include Dr. Charles Shelton, Shell Research, Ltd.; Dr. Ernesto Cespedes, USAE WES; Mr. Robert Hastings, Shell Westhollow Technology Center; Dr. Steve Lieberman, NRAD; and Mr.William Davis, USAE WES. Primary authors of the monograph are Dr. Kenny and Ms. Pepper (Tufts), Mr. Schwartz (ENSR), and Dr. Shelton (Shell). The senior monograph editor is Dr. Balshaw-Biddle (Rice). Dr. Matthew Fraser and Dr. Frank Tittel (both Rice) provided valuable assistance with technical editing.

Additional information about this research can be requested from the principal investigator, Dr. Jonathan E. Kenny (Tufts University, Department of Chemistry, 62 Talbot Ave., Medford, MA 02155) or the AATDF director, Dr. C. Herb Ward (Rice University, EESI MS-316, 6100 Main St., Houston, TX 77005-1892).

AATDF Monographs

This monograph is one of a ten-volume series that record the results of the AATDF Program:

- Surfactants and Cosolvents for NAPL Remediation (A Technology Practices Manual)
- Sequenced Reactive Barriers for Groundwater Remediation
- Modular Remediation Testing System
- Phytoremediation of Hydrocarbon-Contaminated Soil
- Steam Remediation of Contaminated Soils
- Soil Vapor Extraction: Radio Frequency Heating
- Subsurface Contamination Monitoring Using Laser Fluoresence
- Reuse of Surfactants and Cosolvents for NAPL Remediation
- Remediation of Firing-Range Impact Berms
- Surfactants, Foams and Micoremulsions for NAPL Remediation

Advanced Applied Technology Demonstration Facility
(AATDF)
Energy and Environmental Systems Institute MS-316
Rice University
6100 Main Street
Houston, TX 77005-1892

U.S. Army Engineer
Waterways Experiment Station
3909 Halls Ferry Road
Vicksburg, MS 39180-6199

Rice University
6100 Main Street
Houston, TX 77005-1892

Preface

Following a national competition, the Department of Defense (DOD) awarded a $19.3 million grant to a university consortium of environmental research centers led by Rice University and directed by Dr. C. Herb Ward, Foyt Family Chair of Engineering. The DOD Advanced Applied Technology Demonstration Facility (AATDF) Program for Environmental Remediation Technologies was established on May 1, 1993, to enhance the development of innovative remediation technologies for DOD by facilitating the process from academic research to full-scale utilization. The AATDF's focus is to select, test, and document performance of innovative environmental technologies for the remediation of DOD sites.

Participating universities include Stanford University, The University of Texas at Austin, Rice University, Lamar University, University of Waterloo, and Louisiana State University. The directors of the environmental research centers at these universities serve as the Technology Advisory Board (TAB). The U.S. Army Engineer Waterways Experiment Station manages the AATDF Grant for DOD. Dr. John Keeley is the Technical Grant Officer. The DOD/AATDF is supported by five leading consulting engineering firms: Remediation Technologies, Inc., Battelle Memorial Institute, GeoTrans, Inc., Arcadis Geraghty and Miller, Inc., and Groundwater Services, Inc., along with advisory groups from the DOD, industry, and commercialization interests.

Starting with 170 preproposals that were submitted in response to a broadly disseminated announcement, 12 projects were chosen by a peer-review process for field demonstrations. The technologies chosen were targeted at DOD's most serious problems of soil and groundwater contamination. The primary objective was to provide more cost-effective solutions, preferably using *in situ* treatment. Eight projects were led by university researchers, two projects were managed by government agencies, and two others were conducted by engineering companies. Engineering partners were paired with the academic teams to provide field demonstration experience. Technology experts helped guide each project.

DOD sites were evaluated for their potential to support quantitative technology demonstrations. More than 75 sites were evaluated to match test sites to technologies. Following the development of detailed work plans, carefully monitored field tests were conducted, and the performance and economics of each technology were evaluated.

One AATDF project designed and developed two portable Experimental Controlled Release Systems (ECRS) for testing and field simulations of emerging remediation concepts and technologies. The ECRS is modular and portable and allows researchers, at their sites, to safely simulate contaminant releases and study remediation techniques without contaminant loss to the environment. The completely contained system allows for accurate material and energy balances.

The results of the DOD/AATDF Program provide DOD and others with detailed performance and cost data for a number of emerging, field-tested technologies. The program also provides information on the niches and limitations of the technologies to allow for more informed selection of remedial solutions for environmental cleanup.

The AATDF Program can be contacted at Energy and Environmental Systems Institute, MS-316, Rice University, 6100 Main, Houston, TX., 77005, phone 713-527-4700; fax 713-285-5948; or e-mail <eesi@rice.edu>.

The DOD/AATDF Program staff includes

Director:
Dr. C. Herb Ward
Program Manager:
Dr. Carroll L. Oubre
Assistant Program Manager:
Dr. Kathy Balshaw-Biddle
Assistant Program Manager:
Dr. Stephanie Fiorenza
Assistant Program Manager:
Dr. Donald F. Lowe
Financial/Data Manager:
Mr. Robert M. Dawson
Publications Coordinator/Graphic Designer:
Ms. Mary Cormier
Meeting Coordinator:
Ms. Susie Spicer

This volume, *Subsurface Contamination Monitoring Using Laser Fluorescence*, is one of a ten-monograph series that records the results of the DOD/AATDF environmental technology demonstrations. Many have contributed to the success of the AATDF program and to the knowledge gained. We trust that our efforts to fully disclose and record our findings will contribute valuable lessons learned and help further innovative technology development for environmental cleanup.

Katherine Balshaw-Biddle
Carroll L. Oubre
C. Herb Ward

Authors and Editors

Jonathan E. Kenny

Jonathan E. Kenny received a bachelor's degree in chemistry from the University of Notre Dame and master's and Ph.D. degrees from the University of Chicago. His doctoral thesis describes the spectroscopy and dynamics of van der Waals clusters of iodine with helium, neon, hydrogen, and deuterium, studied by laser-induced fluorescence in a supersonic jet expansion. He did a postdoctoral fellowship with Professor Bryan Kohler at Wesleyan University, where he began studying the spectroscopy and photophysics of organic molecules.

As a faculty member in the Department of Chemistry at Tufts University in Medford, Massachusetts, since 1981, he continued his study of photophysics, concentrating on aromatic molecules. Since 1983, he has also been developing new analytical tools for in situ monitoring of soil and ground water contaminants, based on fluorescence and Raman spectroscopy. His research in this area has been described in feature articles in *Spectroscopy*, *Photonics Spectra*, and the newsletters of the MIT Spectroscopy Laboratory and Continuum Laser Corporation. He has also published several articles on the global warming potential of CFC substitutes. Since 1995, he has been a member of the faculty board of the Center for Field Analytical Studies and Technology at Tufts University, which is also an NSF Center of Excellence. He is a member of the editorial advisory board of *Advances in Environmental Research.* He has been involved in many environmental education initiatives, including the Tufts Environmental Literacy Institute (for college and university faculty of all disciplines), Environmental Chemistry for nonscientists, and incorporation of environmental themes into the General Chemistry curriculum. He also teaches in the American Studies program at Tufts.

Jane Wu Pepper

Jane Wu Pepper is a Ph.D. candidate in the Department of Chemistry, Tufts University, where she is doing research under the direction of Professor Jonathan E. Kenny. She has an M.S. degree in analytical chemistry from the State University of New York at Oswego and a B.S. degree in chemistry from Fudan University, Shanghai, China. Ms. Pepper had worked on several environmental projects prior to entering Tufts, including studies on the degradation of polychlorinated biphenyls through photochemical and biological pathways. Recently she has been working on the design, development, and implementation of multiple-channel optical fiber instrumentation deployed with a cone penetrometer for in situ laser-induced fluorescence detection. Along with her colleagues, she was actively involved in field tests of the instrument at Hanscom Air Force Base and Otis Air National Guard Base in Massachusetts. She presented her results at the 1997 FACSS conference (Providence, RI) and at the 1998 SPIE conference (Boston, MA).

Andrew O. Wright

Andrew O. Wright earned his B.A. degree in chemistry from Randolph-Macon College in Ashland, Virginia. He worked as a chemical patent searcher with Cushman, Darby, and Cushman for two years and then enrolled in the doctoral program in chemistry at the University of Wisconsin, where he served as a teaching and research assistant. He received his Ph.D. in chemistry–analytical sciences in 1994. He performed his doctoral research, "Solid State Defect Dynamics Measured with Site Selective Spectroscopy," under the direction of Professor John C. Wright. He was a Postdoctoral Fellow with Dr. Michael Seltzer at the Naval Air Warfare Center, Weapons Division, China Lake, California, before coming to Tufts University in August 1995 to coordinate the construction, testing, and field demonstration of the LIF–EEM system for subsurface contaminant detection. Since completion of the project, he has taken a position with Delta F. Corporation in Woburn, Massachusetts.

Yu-Min Chen

Yu-Min Chen earned a B.S. in chemistry from Shaanxi Teacher's University in Xi'an, China; he later won a Teaching Research Award from the Shaanxi Provincial High Education Bureau. He earned an M.A. degree in chemistry from the University of Northern Iowa and a Ph.D. in chemistry from the University of Utah. His doctoral research, under the direction of Professor Peter B. Armentrout, involved the study of gas-phase reactions of transition-metal cations with hydrocarbon molecules. He won the Research Presentation Award of the Department of Chemistry at Utah in 1992. In 1995 he came to the Boston area as a postdoctoral fellow in the laboratories of Nobel laureate Mario Molina. He was involved in the development and field demonstration of both the EPA- and the DOD-funded prototype LIF–EEM instruments as a postdoctoral fellow in Professor Kenny's laboratory. In particular, he developed the factor analysis methods used to extract chemical speciation and quantitation information from the field data. He has since taken a research position with Acton Research Corporation in Acton, Massachusetts.

Steven L. Schwartz

Steven L. Schwartz is a senior program manager for ENSR Corporation, a national and international environmental consulting firm. His work involves the application of innovative site characterization techniques to industrial soil and ground water quality concerns, where he provides technical leadership and quality control to groups of multidisciplinary professionals involved in site assessment and remediation. During the past 17 years, he has completed projects in a variety of industrial and geologic settings and under diverse state and federal regulatory programs.

Mr. Schwartz was the U.S. representative for the Technology Development Committee of Fugro Environmental (recently acquired by ENSR), which identified and funded new environmental technologies compatible with cone penetrometer testing equipment. He holds a B.S. degree in geosciences from the University of Arizona and is a registered professional geologist.

Charles G. Shelton

Charles G. Shelton is an environmental scientist and technical service consultant with the Shell Global Solutions HSE Contaminated Land Group at Shell Research and Technology Centre, U.K. He obtained both his First Degree and Doctorate in Material Science from the University of Cambridge, U.K., before joining Shell in 1986. At Shell, he applies risk-based approaches to issues regarding management of hydrocarbon and petrochemical contamination in Europe, the Middle East, Africa, and the Americas. In this capacity Shelton identifies, develops, and applies appropriate technologies for rapid, cost-effective assessment of contaminated land. He collaborated with the authors of this monograph during an assignment with the Shell Oil Company Environmental Directorate at the Westhollow Technology Center in Houston, Texas.

Katherine Balshaw-Biddle

Katherine Balshaw-Biddle is an Assistant Program Manager with AATDF at Rice University where she manages three innovative technology projects including creation of the ECRS modular testing unit for remediation technologies. Dr. Balshaw-Biddle has a Ph.D. in geology from the Rice University and a B.S. and M.S. in geology from Michigan State University.

In her capacity as project manager for AATDF, Dr. Balshaw-Biddle provides managerial guidance and technical expertise for the organization, implementation and field demonstration of several projects. She is also an active participant in preparation of reports for each project.

Prior to joining the AATDF, Dr. Balshaw-Biddle worked as a senior geologist for Exxon Production Research Co. and Law Engineering and as a research scientist at Rice University, Department of Environmental Science and Engineering. She has several publications related to environmental technologies, remediation, and sedimentology.

Carroll L. Oubre

Carroll L. Oubre is the Program Manager for the DOD/AATDF Program. As Program Manager he is responsible for the day-to-day management of the $19.3 million DOD/AATDF Program. This includes guidance of the AATDF staff, overview of the 12 demonstration projects, assuring project milestones were met within budget, and that complete reporting of the results are timely.

Dr. Oubre has a B.S. in chemical engineering from the University of Southwestern Louisiana, an M.S. in chemical engineering from Ohio State University, and a Ph.D. in chemical engineering from Rice University. He worked for Shell Oil Company for 28 years; his last job was Manager of Environmental Research and Development for Royal Dutch Shell in England. Prior to that, he was Director of Environmental Research and Development at Shell Development Company in Houston, Texas.

C. H. (Herb) Ward

C. Herb Ward is the Foyt Family Chair of Engineering in the George R. Brown School of Engineering at Rice University. He is also Professor of Environmental Science and Engineering, and Ecology and Evolutionary Biology.

Dr. Ward has undergraduate (B.S.) and graduate (M.S. and Ph.D.) degrees from New Mexico State University and Cornell University, respectively. He also earned the M.P.H. in environmental health from the University of Texas.

Following 22 years as Chair of the Department of Environmental Science and Engineering at Rice University, Dr. Ward is now Director of the Energy and Environmental Systems Institute (EESI), a university-wide program designed to mobilize industry, government, and academia to focus on problems related to energy production and environmental protection.

Dr. Ward is also Director of the Department of Defense, Advanced Applied Technology Demonstration Facility (AATDF) Program, a distinguished consortium of university-based environmental research centers supported by consulting environmental engineering firms to guide selection, development, demonstration, and commercialization of advanced applied environmental restoration technologies for the DOD. For the past 18 years, he has directed the activities of the National Center for Ground Water Research (NCGWR), a consortium of universities charged with conducting long-range exploratory research to help anticipate and solve the nation's emerging groundwater problems. He is also Co-Director of the EPA-sponsored Hazardous Substances Research Center/South & Southwest (HSRC/S&SW), whose research focus is on contaminated sediments and dredged materials.

Dr. Ward has served as President of both the American Institute of Biological Sciences and the Society for Industrial Microbiology. He is the founding and current Editor-in-Chief of the international journal *Environmental Toxicology and Chemistry.*

AATDF Advisors

University Environmental Research Centers

National Center for Ground Water Research
Dr. C. H. Ward
Rice University, Houston, TX

Hazardous Substances Research Center–South and Southwest
Dr. Danny Reible and Dr. Louis Thibodeaux
Louisiana State University, Baton Rouge, LA

Waterloo Centre for Groundwater Research
Dr. John Cherry and Mr. David Smyth
University of Waterloo, Ontario, Canada

Western Region Hazardous Substances Research Center
Dr. Perry McCarty
Stanford University, Stanford, CA

Gulf Coast Hazardous Substances Research Center
Dr. Jack Hopper and Dr. Alan Ford
Lamar University, Beaumont, TX

Environmental Solutions Program
Dr. Raymond C. Loehr
University of Texas, Austin, TX

DOD/Advisory Committee

Dr. John Keeley, Co-Chair
Assistant Director, Environmental Laboratory
U.S. Army Corps of Engineers
Waterways Experiment Station, Vicksburg, MS

Mr. James I. Arnold, Co-Chair
Acting Division Chief, Technical Support
U.S. Army Environmental Center
Aberdeen, MD

Dr. John M. Cullinane
Program Manager, Installation Restoration
U.S. Army Corps of Engineers, Vicksburg, MS
Waterways Experiment Station
Vicksburg, MS

DOD/Advisory Committee (continued)

Mr. Scott Markert and Dr. Shun Ling
Naval Facilities Engineering Center
Alexandria, VA

Dr. Jimmy Cornette, Dr. Michael Katona and Major Mark Smith
Environics Directorate
Armstrong Laboratory
Tyndall AFB, FL

Commercialization and Technology Transfer Advisory Committee

Mr. Benjamin Bailar, Chair
Dean, Jones Graduate School of Administration
Rice University, Houston, TX

Dr. James H. Johnson, Jr., Associate Chair
Dean of Engineering
Howard University, Washington, DC

Dr. Corale L. Brierley
Consultant
VistaTech Partnership, Ltd.,
Salt Lake City, UT

Dr. Walter Kovalick
Director, Technology Innovation Office
Office of Solid Wastes and Emergency Response
U.S. EPA, Washington, DC

Mr. M. R. (Dick) Scalf (retired)
U.S. EPA
Robert S. Kerr Environmental Research Laboratory
Ada, OK

Mr. Terry A. Young
Executive Director
Technology Licensing Office
Texas A&M University, College Station, TX

Mr. Stephen J. Banks
President
BCM Technologies, Inc., Houston, TX

Consulting Engineering Partners

Remediation Technologies, Inc.
Dr. Robert W. Dunlap, Chair
President and CEO
Concord, MA

Parsons Engineering
Dr. Robert E. Hinchee (Originally with Battelle Memorial Institute)
Research Leader
South Jordan, UT

GeoTrans, Inc.
Dr. James W. Mercer
President and Principal Scientist
Sterling, VA

Arcadis Geraghty & Miller, Inc.
Mr. Nicholas Valkenburg and Mr. David Miller
Vice Presidents
Plainview, NY

Groundwater Services, Inc.
Dr. Charles J. Newell
Vice President
Houston, TX

Industrial Advisory Committee

Mr. Richard A. Conway, Chair
Senior Corporate Fellow
Union Carbide, S. Charleston, WV

Dr. Ishwar Murarka, Associate Chair
Electric Power Research Institute
Currently with Ish, Inc., Cupertino, CA

Industrial Advisory Committee (continued)

Dr. Philip H. Brodsky
Director, Research and Environmental Technology
Monsanto Company, St. Louis, MO

Dr. David E. Ellis
Bioremediation Technology Manager
DuPont Chemicals, Wilmington, DE

Dr. Paul C. Johnson
Department of Civil Engineering
Arizona State University, Tempe, AZ

Dr. Bruce Krewinghaus
Shell Development Company
Houston, TX

Dr. Frederick G. Pohland
Department of Civil and Environmental Engineering
University of Pittsburgh, Pittsburgh, PA

Dr. Edward F. Neuhauser, Consultant
Niagara Mohawk Power Corporation
Syracuse, NY

Dr. Arthur Otermat, Consultant
Shell Development Company
Houston, TX

Mr. Michael S. Parr, Consultant
DuPont Chemicals
Wilmington, DE

Mr. Walter Simons, Consultant
Atlantic Richfield Company
Los Angeles, CA

Acronyms and Abbreviations

AATDF	Advanced Applied Technology Development Facility
AFA	abstract factor analysis
AFB	U.S. Air Force Base
ANGB	U.S. Air National Guard Base
BTEX	benzene, toluene (methylbenzene), ethylbenzene, and xylene (dimethylbenzene; includes three structural isomers: *o*-, *m*-, *p*-xylene)
CCD	charge-coupled device
CPT	cone penetrometer technology
DOD	U.S. Department of Defense
DNAPLs	dense nonaqueous phase liquids
DRO	diesel range organics
EEM	excitation–emission matrix
EPA	U.S. Environmental Protection Agency
FVD	fluorescence versus depth log
GC/MS	gas chromatography/mass spectrometry
HG	harmonic generator
HPLC	high-performance liquid chromatography
LIF	laser-induced fluorescence
NA	numerical aperture
NAPL	nonaqueous phase liquid
Nd:YAG	neodymium:yttrium aluminum garnet
NJ	nannoJoule
NRAD	Naval Research and Development
PAH	polynuclear aromatic hydrocarbon (or polyaromatic hydrocarbons)
PDA	photodiode array
PMT	photomultiplier tube
POL	petroleum, oils, and lubricant
RAFA	rank annihilation factor analysis
RhB	rhodamine B
ROST	Rapid Optical Screening Tool
RSD	relative standard deviation
SCAPS	Tri-Service Characterization and Analysis Penetrometer System
SRS	stimulated Raman scattering
TDGC/MS	thermal desorption, gas chromatography/mass spectrometry
TFA	target factor analysis
TIR	total internal reflection
TPH	total petroleum hydrocarbon
TPHCWG	Total Petroleum Hydrocarbon Criteria Working Group
WES	U.S. Army Engineer Waterways Experiment Station
WTM	wavelength time matrix

Contents

Executive Summary

A new laser-induced fluorescence (LIF) probe developed by researchers at Tufts University identifies and semiquantifies classes of hydrocarbon compounds in contaminated subsurface soils. The probe is advanced into the subsurface using cone penetrometer testing (CPT) equipment and identifies hydrocarbon classes by use of a multichannel excitation–emission matrix (EEM). The LIF-EEM instrumentation is capable of assessing common subsurface hydrocarbon contaminants at depths up to 50 ft. The technique allows for the collection of significant amounts of subsurface information, surpassing conventional data collection methods that can be used to rapidly identify areas of concern beneath a site. The technology has significant application for environmental assessment and remediation programs and has potential applications for monitoring various manufacturing processes and industrial wastewater operations.

This monograph describes the development, testing, and performance of the Tuft's LIF-EEM instrumentation and summarizes applicability of the technology. Funding for the project was provided by the U.S. Department of Defense to the Advanced Applied Technology Demonstration Facility (AATDF) at Rice University in Houston, Texas. The work is aimed at commercialization of LIF-EEM technology.

The ideas behind the design of the LIF-EEM probe are discussed in detail in a design manual, which was prepared as part of the AATDF-sponsored program and is attached to this monograph as Appendix A. A manual describing the assembly, calibration, and operation of the probe with CPT direct-push equipment was also prepared and is included here as Appendix B. This monograph will be of interest to parties who are considering further development, commercialization, or licensing of a LIF probe.

The technology performance evaluation presented here focuses on three major project objectives to

- Document the ability of LIF intensity and EEM data to semiquantify classes of hydrocarbon compounds in soil samples during laboratory and field measurements
- Demonstrate advantages of an LIF probe that uses simultaneous multiple wavelength excitation
- Validate the use of multivariate statistical analysis of EEMs to classify and semiquantify hydrocarbons

Field test results of the LIF-EEM instrumentation provide reasonable grounds for optimism that the system will be capable of providing in situ, semiquantitative hydrocarbon classifications. Promising linear correlations of LIF-EEM data and analytical measurements (total petroleum hydrocarbons, diesel range organics, and naphthalene) have been observed in a limited number of soil types from two test sites. The use of multivariate statistical methods to classify and semiquantify groups of aromatic compounds also appears promising, although test results were limited to the semivolatile compounds that were present in the tested soil samples.

The LIF-EEM techniques show considerable merit, particularly the combination of hydrocarbon classification and semiquantification. It is probably more realistic to expect success with the classification technique, especially during analysis of complex contaminant mixtures rather than identification of individual constituents, although this has been accomplished for a limited number of samples in the laboratory. Further effort to identify the role and effects of subsurface materials on quantification and estimation of detection limits is being addressed as well as additional work to enhance the LIF system so that the volatile BTEX constituents can be detected and quantified in the field. The current instrumentation has not demonstrated the sensitivity to detect and quantify these light aromatic constituents. These compounds are often the constituents of concern (COCs) during human health risk analyses, and identification of these COCs is important to accurately complete a site characterization. Additional effort is also being undertaken to fully automate data collection, treatment, and reporting methods and to allow collection of data without pausing advancement of the CPT probe.

The LIF-EEM probe on CPT equipment is a screening tool for site characterization and could provide an economical, rapid assessment of contaminated sites. Data resolution on a cm scale coupled with the significant volume of subsurface information generated by this technique provide an excellent database to use with commercially available software for visualizing the three-dimensional extent of contaminants.

Use of the instrument is dependent on its ability to penetrate subsurface soils to the desired investigation depth. Consequently, the probe may not be useful in areas of crystalline rock, heavily cemented sediments or locations with thick zones of coarse gravel or large cobbles although the probe could be lowered down a previously cut uncased borehole. Under favorable soft-soil conditions, the LIF-CPT probe can be easily advanced to its current design limit of 50 ft.

Whereas existing LIF systems may help ease the introduction of EEM technology into the commercial market, these same systems also present significant competition as their developers continue to improve on equipment capabilities and market presence. Competing technologies capable of achieving similar site characterization objectives are also being commercialized by researchers at national laboratories. Consequently, there is a current window of opportunity for introduction of this EEM technology.

CHAPTER 1

Introduction

1.1 LIF-CPT TECHNOLOGY FOR SITE MONITORING

Laser-induced fluorescence (LIF) is a technique that provides considerable sensitivity in the detection of fluorescent analytes, based upon laboratory experiments of single-molecule detection limits.[1-3] In the realm of geophysical testing of soils and subsurface environments, cone penetrometer testing (CPT) has proven its effectiveness. In the past decade the potential of CPT for delivering chemical probes to the subsurface has been recognized and developed, and the combination of LIF and CPT has been particularly successful.[4-9] The heart of the CPT system is a cone-tipped rod that is pushed into the ground using a hydraulic system mounted on a truck or trailer. Advantages of using the CPT system for advancement of the LIF probe include the following:

- During a push, or sounding, CPT equipment can advance the LIF probe at a controlled rate and identify soil conditions in the same interval where the LIF probe collects fluorescence data.
- LIF data collection systems are housed inside the cone truck, allowing work to proceed in adverse weather without affecting data quality.
- The cone penetrometer direct push approach reduces potential exposure of site personnel to contaminants.
- There are no soil cuttings generated during LIF equipment advancement and, consequently, minimal potentially hazardous materials to dispose.

Compared to conventional soil sampling and laboratory analyses, LIF-CPT has proven to be a rapid, cost-effective approach to environmental site characterization and monitoring. It provides real-time data collection for mapping contaminant distribution in the subsurface, which surpasses conventional data collection methods.

Current LIF-CPT techniques, described in Chapter 7 of this monograph, typically focus on single-wavelength excitation, without distinction of the full spectral "fingerprints" of different fluorescing species.[10] This monograph describes the development, testing, performance, and applicability of a new LIF probe that identifies and semiquantifies classes of hydrocarbon compounds in contaminated soils. The probe uses simultaneous multiple-wavelength excitation and emission detection and processes the data as an excitation–emission matrix (EEM). Currently, LIF instrumentation can provide in situ, real-time data to assess common subsurface hydrocarbon contaminants at concentrations in the parts per million (ppm) level at depths up to 50 ft or more (determined by the length of the optical cable).

The Tufts' LIF-EEM probe and instrumentation have been integrated with a commercial CPT vehicle, field tested, and demonstrated at Hanscom Air Force Base and Otis Air National Guard Base in Massachusetts. Qualitative and semiquantitative chemical data were derived from in situ hydrocarbon contaminants during these tests. These data, integrated with the CPT depth and lithologic data, were used to complete site characterizations of the test areas at each base.

This equipment was designed and assembled at Tufts University, in Medford, Massachusetts, by the research team of Dr. Jonathan E. Kenny. Funding for this project was provided by the U.S. Department of Defense to the Advanced Applied Technology Demonstration Facility (AATDF) headquartered at Rice University in Houston, Texas. Cost sharing included support from Perkin-Elmer Corporation, SLM/Aminco, and Tufts University. This work tests the viability and applicability of the Tufts LIF-EEM system, which has proven potential in environmental assessment and site remediation, and significant potential in monitoring manufacturing processes and industrial wastewater operations. This monograph will be of interest to parties who are concerned with environmental or industrial applications of LIF or are considering commercialization or licensing of the probe and the LIF-EEM technology.

Earlier publications produced during this project include the Work Plan, Work Plan Addendum, Design Manual, Calibration and Operations Manual, Final Technical Report, and the Technology Evaluation Report. This monograph draws on information from all six reports in addition to comments derived from the AATDF peer review process and the project technical advisors (shepherds).

1.2 PROJECT BACKGROUND AND PHASES

Two generations of the LIF-EEM tool have been designed, assembled and tested at Tufts University. The first instrument[9, 11-13] assembled onto a cone penetrometer had 10 sapphire windows aligned 1.5 in apart along the probe's vertical axis. Each window was fitted with one pair of excitation–emission fibers. This instrument was integrated with a CPT truck similar to that used in the Tri-Service Characterization and Analysis Penetrometer System (SCAPS). The tool was used in projects funded by the U.S. Environmental Protection Agency (EPA) for assessment of contamination at Hill Air Force Base (AFB) in Utah,[11] Hanscom AFB in Massachusetts,[9,13,14] and the Elizabeth City U.S. Coast Guard Station in North Carolina.[14]

To improve capabilities of the EPA-funded tool, a second-generation LIF system (the subject of this monograph) was constructed at Tufts under the AATDF program. The hardware used in the current LIF-EEM instrument incorporates several changes and improvements over the first-generation system. The probe was redesigned to allow all 10 excitation–emission fiber pairs to send and receive signals through one 1-cm-diameter sapphire window. This modification was made to eliminate the need for depth correction of fluorescence data collected from separate windows. The footprint of the combined laser, Raman shifter, and associated optics was also reduced from the original 36 × 60 in to about 24 × 48 in, making mobilization and integration with the CPT truck more efficient. In addition, the repetition rate for the Nd:YAG laser was upgraded from 20 to 50 pulses per second, or Hertz (Hz), which permits faster acquisition of fluorescence data.

This second-generation LIF-EEM tool was assembled and installed in a CPT truck provided by Fugro Geosciences, Inc. The probe was tested and calibrated at Tufts University and then deployed to characterize subsurface soil contamination at Hanscom AFB near Bedford, Massachusetts, and at Otis Air National Guard Base (ANGB) in Falmouth, Massachusetts. Results of these tests are detailed in an AATDF Final Technical Report written by Dr. Kenny's research team. An evaluation of the system performance is discussed in an AATDF Technology Evaluation Report prepared by the project's engineering consultants, ENSR Corporation, with data evaluation performed by Shell Research, Ltd.

The project was approached in four phases, with specific tasks and deliverables for each phase. Details on the phases were presented in an AATDF Work Plan and Work Plan Addendum and are summarized here.

1. System design; selection and purchase of major system components; assembly and testing of instrumentation; development of standard operating procedures for data collection; completion of LIF laboratory tests
2. Fabrication, assembly, and calibration of the LIF probe components; integration of the probe with CPT equipment; execution of Hanscom AFB field test.
3. Performance evaluation of LIF probe, the system, data handling techniques, and data processing software from the Hanscom test; preparation of AATDF Calibration and Operations Manual; execution of Otis ANGB field test.
4. Performance evaluation of LIF probe, the system, data handling techniques and data processing software from the Otis ANGB field test; submittal of four AATDF reports (Design Manual, Calibration and Operations Manual, Site Report, and the Final Technical Report).

CHAPTER 2

LIF-EEM Engineering Design

Use of laser-induced fluorescence (LIF) technology for in situ contaminant monitoring is based upon the principle that many petroleum hydrocarbons absorb laser energy and quickly release the excess energy by emitting light, i.e., fluorescing. The LIF-excitation–emission matrix (EEM) system generates the laser light, delivers it to the remote sample using fiber optics, collects and delivers the resulting emission from the fluorescing hydrocarbons to a detector, and records the signal electronically. The Tufts system design advances the state of the art of in situ LIF by delivering multiple excitation wavelengths and detecting fluorescence as a function of both excitation and emission wavelengths to create a three-dimensional fingerprint of the contaminant. This can be accomplished in situ during a CPT sounding, thus providing real-time measurements of contaminant distribution.

2.1 EXCITATION–EMISSION MATRIX

The multiple excitation laser wavelengths are generated simultaneously by means of stimulated Raman scattering (SRS). In this process, incoming laser light of a single wavelength is coherently and inelastically scattered by the active medium (in this system, a compressed gas) to generate an outgoing laser beam containing a range of new wavelengths. The new wavelengths are shifted from the wavelength of the incoming beam by amounts that depend on the vibrational frequencies of the molecules that comprise the active medium. Therefore, the values of the output wavelengths may be selected by choosing the incoming wavelength and the identity of the Raman-scattering gas appropriately.

The multiple emission wavelengths are monitored by means of a grating spectrograph equipped with a charge-coupled device (CCD) detector. This spectroscopic instrumentation collects and displays fluorescence signals from the sample in real time as the probe is advanced. The signal is mathematically manipulated as a matrix (the EEM) of intensities whose row and column indices are the excitation and emission wavelengths. Mathematical routines and calibration standards are used for extracting information from this matrix to identify and quantify the chemical composition of hydrocarbon contaminants in the soil.

The EEMs provide the potential for delineating the distribution of hydrocarbon contaminants in soil and for distinguishing particular hydrocarbon compounds or classes of hydrocarbon compounds within the affected area. These data are recorded simultaneously with the CPT information on depth, soil type, and other physical soil parameters.

2.2 DESIGN GOALS

Specific goals were adopted for design of the LIF-EEM tool to advance its capacity beyond current LIF-CPT instrumentation, including Fugro Geosciences' Rapid Optical Screening Tool (ROST™), the Tri-Service Site Characterization and Analysis Penetrometer System (SCAPS), and the first generation LIF-EEM instrumentation developed by researchers at Tufts University for the U.S. Environmental Protection Agency (EPA).

The principal design goals for the new LIF instrument included the following:

- Expand the number of excitation wavelengths to broaden the sensing capability of the instrument, improve contaminant detection limits to enhance the system's ability to discriminate among polycyclic aromatic hydrocarbons (PAHs), and provide sensitive detection of single-ring aromatics (particularly benzene toluene, ethylbenzene, and xylenes, or BTEX),
- Demonstrate the value of EEMs in identifying classes of contaminants,
- Streamline data processing efforts and user interface during data collection,
- Reduce the size of the LIF probe and area of the sampling window in the probe,
- Enhance sensitivity of the multiple channel technique through efficient light collection and reduction of noise caused by instrument artifacts,
- Allow real-time and automatic operation of the LIF-CPT probe without stopping a CPT push to collect EEM data,
- Reduce the footprint of the instrument and support equipment associated with efficient deployment in a commercial CPT vehicle.

2.3 SELECTION OF EXCITATION WAVELENGTHS

An important distinction of the LIF-EEM instrumentation compared to other LIF-CPT tools is the use of multiple, simultaneous excitation wavelengths to enable EEM analysis. The innovation of simultaneous delivery of multiple wavelengths comes with additional equipment cost and potential complications, so it is necessary to document the analytical advantage gained as the number of excitation wavelengths is increased.

Tufts University researchers evaluated the use of 5 to 14 excitation wavelengths in the ultraviolet (UV) range to determine how the number of excitation wavelengths statistically affects the ability to identify constituents in a multicomponent mixture of common aromatic compounds.[15] Fourteen were chosen as practical excitation wavelengths given the technical strategy of Raman shifting, available space in cone penetrometer rods for fiber optic cabling, size of fiber, spectrograph and detector, and focusing aberrations of flat-field spectrographs. However, based on the data presented below, 10 wavelengths were ultimately selected as the optimal excitation wavelengths for the field tests.

A mixture consisting of a three-component solution of one-, two-, and three-ring aromatic molecules (benzene, naphthalene, and anthracene) at known concentrations in cyclohexane was tested. Emission spectra were collected on a conventional fluorimeter at 14 equally spaced wavelengths from 245 to 375 nm in the UV region of the spectrum. Similarly, EEMs of three one-component standard solutions of benzene, naphthalene, and anthracene in cyclohexane were measured.

To study the effect of matrix size on the accuracy of the data analysis, the 14-excitation-channel EEM was truncated to create EEMs with 10, 7, and 5 excitation channels. For the smaller numbers of excitation wavelengths, EEMs representing different choices of the excitation wavelengths from the full data set were selected. The first data set consisted of EEMs corresponding to all 14 excitation wavelengths. The second data set involved EEMs for 10 excitation wavelengths (245, 255, 265, 275, 285, 315, 325, 335, 345, and 355 nm) constructed by eliminating the four wavelengths with the lowest corresponding emission signals (295, 305, 365, and 375). The third data set divided the original 14 excitation wavelengths into two sets of seven designated A (255, 275, 295, 315, 335, 355, and 375 nm) and B (245, 265, 285, 305, 325, 345, and 365 nm). Finally, three sets of five excitation wavelengths were chosen and designated A (255, 275, 295, 345, and 355 nm), B (245, 265, 315, 365, and 375 nm), and C (255, 285, 305, 325, and 335 nm). The multicomponent EEMs from each data set were analyzed by least-squares methods.[16] The concentrations obtained for the individual components (benzene, naphthalene, and anthracene) showed close agreement with actual concentrations for the full data set as well as the various subsets (Table 1).

Table 1 Least-Squares Analysis of Three-Component Mixture (Benzene, Naphthalene and Anthracene)

	Estimated Relative[a] Concentration	Standard Deviation	True Relative[a] Concentration	% Error
	14 Excitation EEMs			
Anthracene	0.303	0.145	0.300	+1.00
Naphthalene	0.306	0.056	0.300	+2.10
Benzene	0.272	0.036	0.300	−9.37
	10 Excitation EEMs			
Anthracene	0.303	0.145	0.300	+0.83
Naphthalene	0.314	0.059	0.300	+4.67
Benzene	0.270	0.041	0.300	−9.93
	7 Excitation EEMs (using wavelength set A)			
Anthracene	0.322	0.205	0.300	+7.47
Naphthalene	0.292	0.085	0.300	−2.77
Benzene	0.270	0.045	0.300	−10.00
	7 Excitation EEMs (using wavelength set B)			
Anthracene	0.277	0.201	0.300	−7.77
Naphthalene	0.319	0.074	0.300	+6.30
Benzene	0.275	0.059	0.300	−8.33
	5 Excitation EEMs (using wavelength set A)			
Anthracene	0.322	0.204	0.300	+7.17
Naphthalene	0.306	0.081	0.300	+1.93
Benzene	0.261	0.052	0.300	−13.10
	5 Excitation EEMs (using wavelength set B)			
Anthracene	0.279	0.201	0.300	−7.16
Naphthalene	0.290	0.125	0.300	−3.43
Benzene	0.285	0.053	0.300	−5.10
	5 Excitation EEMs (using wavelength set C)			
Anthracene	0.360	0.189	0.300	+19.97
Naphthalene	0.307	0.083	0.300	+2.33
Benzene	0.247	0.064	0.300	−17.57

[a] Relative concentration = concentration in standard solution/concentration in mixture solution.

Figure 1 shows the average absolute values of percent error and average percent standard deviations as a function of number of excitation wavelengths for all data sets as well as the mean value for each number of excitation wavelengths. It can be seen that percent standard deviation

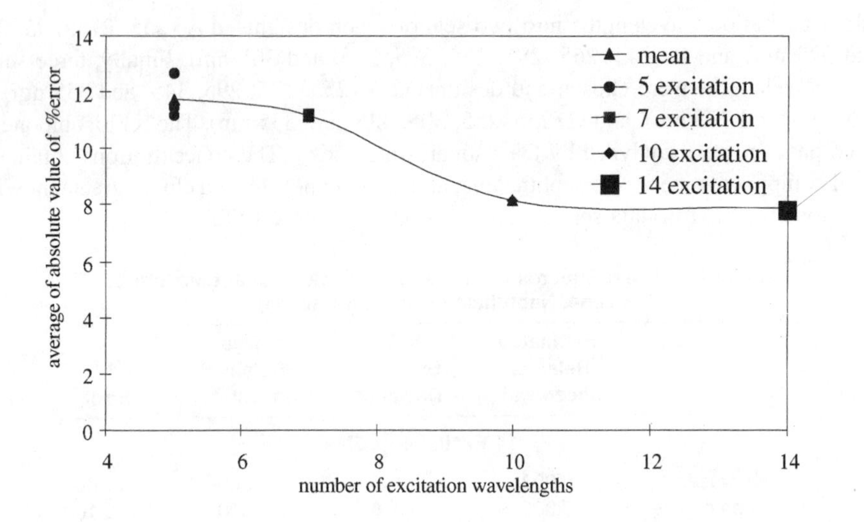

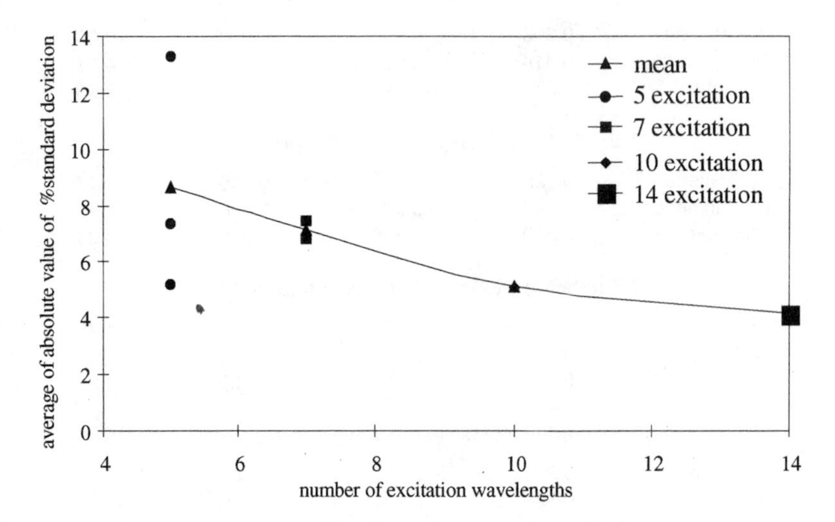

Figure 1 Plot of average of absolute value of percent error and average of the percent standard deviation vs. number of excitation wavelengths used for the 3 component (benzene, naphthalene and anthracene) system.

decreased smoothly as the number of excitation wavelengths increased, whereas the percent error showed a step decrease on going from 7 to 10 excitation wavelengths. For the smaller data sets, the quality of results depends on the exact choice of wavelengths, and such optimization can be achieved with the system if the target analytes are known in advance. Other investigations indicated that additional excitation wavelengths improved classification of compounds, although a similar experiment on a multicomponent mixture consisting of only single-ring aromatics (BTEX) showed that even with the full 14-excitation-wavelength data set it was not always possible to differentiate individual, chemically similar components.[15] As noted, 10 was selected as the optimal number of excitation wavelengths for the field tests, and a strategy was developed to select the specific wavelengths based upon the types of compounds that were expected to be present at the field sites.

2.4 SYSTEM COMPONENTS

The four principal components of the LIF tool are the excitation system (laser and Raman shifter), the optical delivery system (i.e., optical fibers that deliver and collect light), the sample interface (the multichannel probe), and the detection system (CCD/spectrograph). The design, assembly, and initial performance of the components for the Tufts' LIF–CPT are described in the AATDF report "Design Manual for Laser-Induced-Fluorescence Cone Penetrometer Tool," which is included in Appendix A.

2.4.1 Nd:YAG Laser

In the LIF-EEM system, the laser chosen to excite fluorescence is a pulsed, flashlamp-pumped, Nd:YAG (neodymium:yttrium aluminum garnet) unit operating at 1064 nm with a repetition rate of 50 pulses per second. It provides simultaneous and collinear output of two wavelengths, 266.0 and 354.7 nm (Figure 2). The Nd:YAG laser model used in this field work is manufactured by Quantel S.A. It was chosen for its small size, ease of flash lamp replacement, and modest power requirements, which facilitate packaging and field operation; its simultaneous output at two harmonic wavelengths, which permits a greater selection of excitation wavelengths; its high pulse repetition rate, which allows faster acquisition of fluorescence data; and its high peak output power, which increases the efficient generation of Raman-shifted excitation wavelengths.

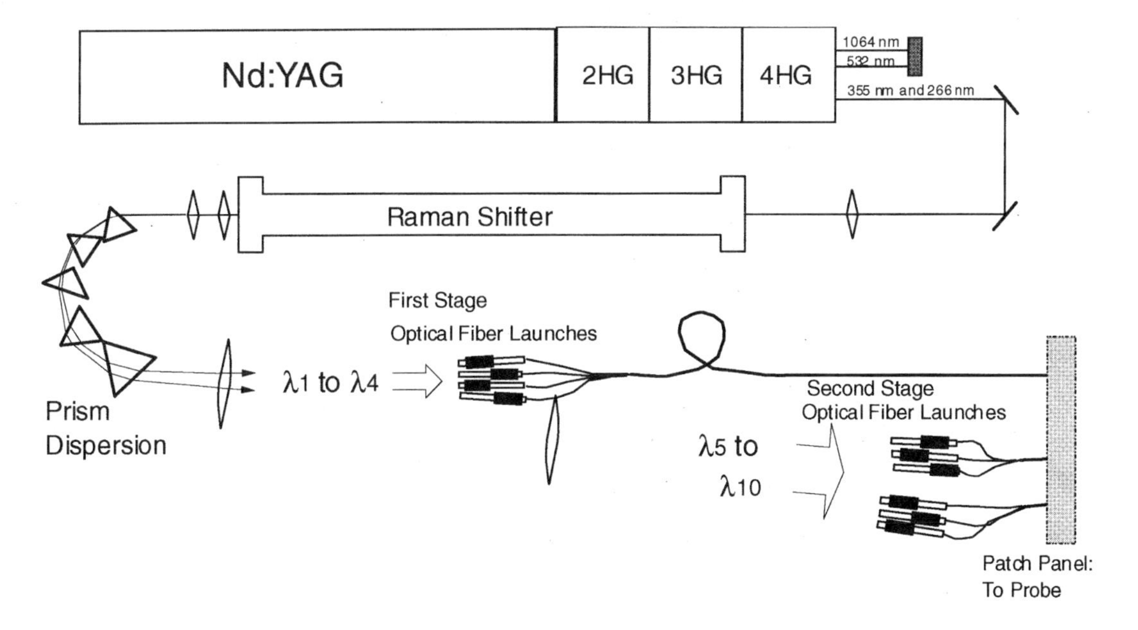

Figure 2 Layout of the optical table used in the current LIF-EEM instrument. Note that only 2 of many output wavelengths are shown, whereas there are actually 10 excitation beams (λ_1 to λ_{10}) that are launched into the optical fibers for LIF measurements.

When both harmonic modules are tuned, 41 to 48% of the total measured power is at 266 nm, and 52 to 59% is at 355 nm. With a fresh flashlamp, average power can reach 700 mW. Output diminishes gradually over the lamp lifetime, which is 40 million shots or continuous operation for 222 hours at 50Hz. When the laser output reaches 400 mW, SRS stability and intensity are no longer acceptable. Average laser power is measured using a thermopile detector and radiometer. It can be converted to pulse energy by dividing the power by the repetition rate of the laser, 50 Hz.

2.4.2 Raman Shifter and Excitation Beams

Fused-silica turning prisms are used to direct laser output into the Raman shifter (Figure 2), a 1-in diameter, 50-cm-long stainless steel cylinder with o-ring sealed fused-silica windows, which contains the gas-active media at pressures up to 250 psi. Valves provide easy gas filling and flushing, and a gas regulator monitors pressure. A blow-off valve releases gas at a preset pressure well below the pressure rating of the windows.

The Raman shifter is placed into two v-groove mounts attached to a precision optical rail so focusing and collimating lenses can be adjusted along an aligned axis. Iris diaphragms define the light path at either end of the optical rail. A 50-cm lens focuses light within the body of the shifter, and a 10/100-mm lens pair forms a telescope to recollimate the output into a 1-in-diameter beam. This beam is directed through a series of five dispersion prisms (Figure 2) that align it through a focusing lens and onto the first stage of the optical fiber launch.

The Raman shifter generates excitation beams at multiple wavelengths using the process of stimulated Raman scattering (SRS). The choice of gas is based on the frequency shift desired and the efficiency, or gain, of the gas at that frequency. H_2 and CH_4 gas are typical SRS gain media. H_2/CH_4 mixed-gas systems have been developed and characterized that produced numerous output beams as a result of the characteristic frequency shifts of H_2 and CH_4 (4155 cm^{-1} and 2916 cm^{-1}, respectively) and their higher order contributions (multiples and linear combinations of these shifts).[17] The Raman shifter used for these LIF measurements employed a 45:55 mixture of H_2/CH_4 at a total pressure of 150 psi at 25°C. The two wavelengths pumping this system, 354.7 and 266.0 nm, created a large number of output lines of varying intensity. The 10 wavelengths selected for excitation of PAHs in soil are summarized in Table 2.

Table 2 SRS Energies Measured through a 4-m-Long Optical Fiber

Wavelength (nm)	SRS Peak Assignment (*P, H, M*)	Measured Energy (μJ)
246.9	4, 0, 1	0.32
257.5	4, 1, –1	0.32
266.0	4, 0, 0	6.8
288.4	4, 0, –1	15.0
299.1	4, -1, 0	26.0
314.8	4, 0, –2	2.6
327.6	4, –1, –1	1.1
341.5	4, –2, 0	1.9
354.7	3, 0, 0	29
416.0	3, –1, 0	6.5

In Table 2, P indicates which harmonic of the Nd:YAG laser is the pump or input beam for a given output beam (4 for fourth harmonic output at 266.0 nm, 3 for third harmonic at 354.7 nm), *H* is the number of hydrogen vibrational quanta associated with the output, and *M* is the number of methane vibrational quanta. Negative values of *H* and *M* correspond to Stokes processes (shifts to a longer wavelength), and positive values are anti-Stokes. The numerical values of *P, H*, and *M* may be inserted into a simple expression to give the frequency of a beam.[17] Wavelengths from other SRS processes are weaker than those mentioned but could be used. These include 239.5 (4, 1, 0), 275.1 (4, –1, 1), 309.1 (3, 1, 0), 321.4 (3, 0, 1), and 395.6 nm (3, 0, –1). For measurement of SRS outputs, each line (after dispersion) was directed into a 4-m optical fiber, and output was measured at the end of the fiber. SRS pulse energies were measured with a RjP-637 energy probe, manufactured by LaserProbe, Utica, New York.

Pulse energy fluctuations are caused by laser output variations due to the following mechanisms:

- *Flashlamp aging*: Aging of the lamp with use affects the fundamental output of the YAG laser at 1064 nm as well as all beams at other wavelengths, which are derived from the fundamental by various nonlinear processes, resulting in generally different amplification of the fluctuations in the fundamental for each beam.
- *Solarization*: Solarization, i.e., defects developed in the optical fibers during transmission of laser output signals (in this case UV wavelengths), occurs relatively quickly in optical fibers, but soon reaches an asymptote, whereupon further degradation occurs very slowly.
- *Alignment operations*:
  1. Tuning of harmonic-generating crystals to optimize conversion efficiency. Crystal tuning affects the relative output of 266 and 355 nm because these harmonics compete for available second harmonic output.
  2. Aligning excitation fibers with focused output beams at the "launch" point where laser beams enter the fibers. Intensity distribution variations occur over a beam cross-section, and if the focused spot diameter nearly equals the fiber diameter (a condition which is usually chosen to minimize power damage to the fiber), then effects of slight misalignment are magnified. Due to mechanical and thermal perturbations such as vibrational CPT motion (travel and tool advancement) and inadequate cooling of the laser power supply, optimization of fiber tip orientation is required. As a result of the alignments of the crystals and the fibers, it is therefore advisable to reestablish system output levels frequently, which can be conducted within 10 minutes.

The stability of SRS beams was measured (without the third harmonic module) over an 8-h period. Figure 3 shows that the beam powers were quite stable. Using both third and fourth harmonic modules, stability is reduced. An example of SRS output energy during an LIF/CPT push is shown in Figure 4.

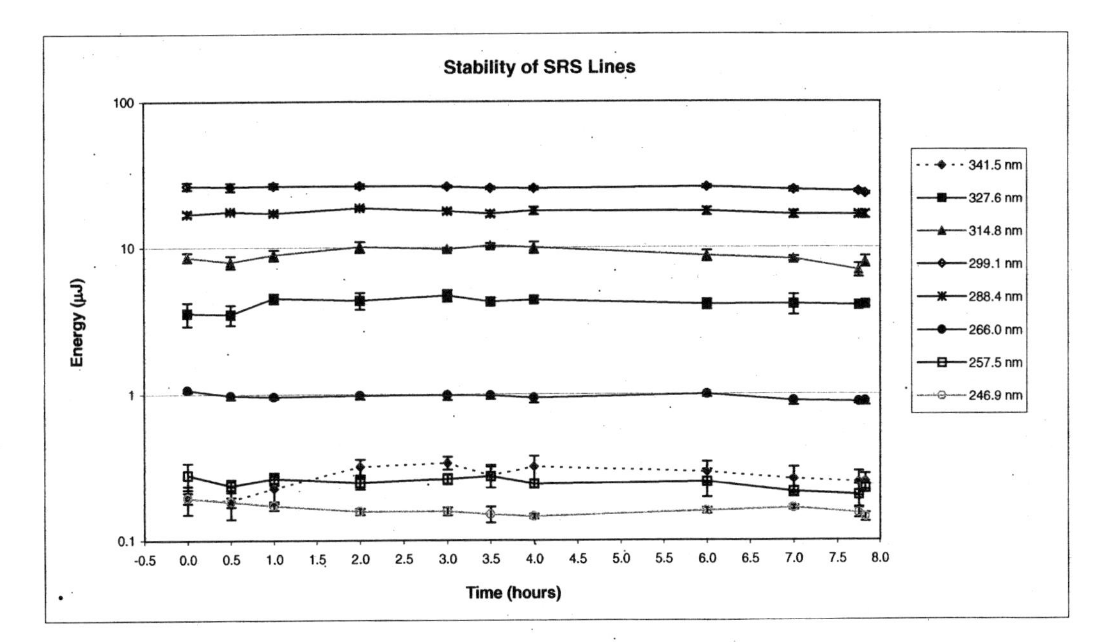

Figure 3 Output energy variation for various SRS excitation wavelengths without third harmonic module.

Monitoring SRS excitation energies at the probe would facilitate quantitative comparison of LIF-EEM data acquired at different times. However, monitoring individual channels with an energy meter is impractical during field operations. Instead, the probe is immersed in a dilute solution of

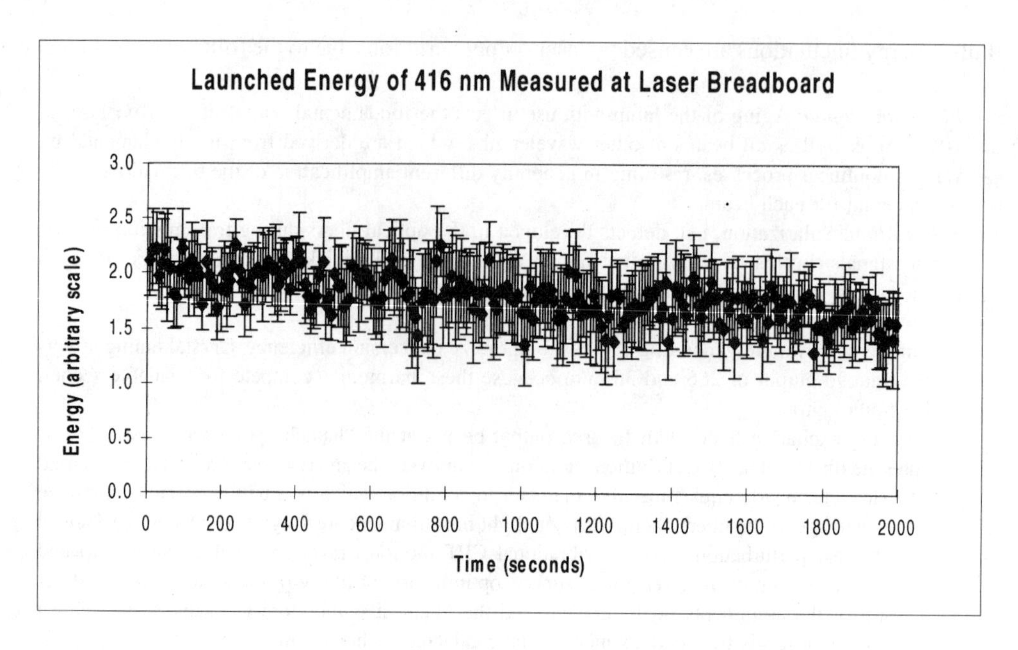

Figure 4 Stability of 416-nm SRS line with second and third harmonic modules.

rhodamine B (RhB) in ethanol (0.2 mg/l) to measure fluorescence before and after an LIF-CPT push. The data are converted to energy or photon intensities using a calibration scheme described in Section 2.5.

2.4.3 Multichannel Optical Fiber and Launch Assemblies

Laser propagation through optical fibers causes some attenuation, but incomplete coupling of SRS light to fibers is also a source of reduced beam output. For the strongest excitation wavelengths (266 and 355 nm and to a lesser extent 288 and 299 nm), the fiber is placed in a defocused region of the laser path to avoid damage.

At the first-stage launch assembly for the optical fibers, the laser beam is focused by a 300-mm lens onto the polished fiber ends positioned in front of each wavelength emanating from the dispersion prisms. Two 150-mm lenses are also positioned side by side to capture the remaining dispersed SRS lines and refocus them into the second stage fiber optic launch (Figure 2).

Each fiber launch assembly consists of a fiber chuck to grip the fiber without stress damage, a miniature translation stage to move the chuck to capture the dispersed laser beam, and reversible aluminum brackets to grip the chuck and fix it to the table. The translation stages are the bulkiest part of the launch assembly, so this part of the design was important in reducing the size of the laser excitation system.

The optical fiber is composed of high-UV-transmitting fused silica with a polyimide buffer layer for flexibility and strength. The numerical aperture of the fiber is 0.22. To provide 12 channels (10 for EEM collection and 2 as spares), each with an excitation and a detection fiber, a total of 24 fibers are attached to the LIF-CPT probe. Six sets of four individual optical fibers encased in cable sheathing are threaded through the hollow 1-in diameter annulus of the CPT rods, in addition to the electrical cable for the CPT geophysical probes (Figure 5). The bundled optical cables have an overall diameter of 0.5 in. Two different optical cable lengths, 20 and 11 m, were assembled for sites with different depth requirements, with an additional backup for each. Because of the fiber attenuation at the shorter wavelengths (below 300 nm), it is desirable not to use a cable much longer than needed.

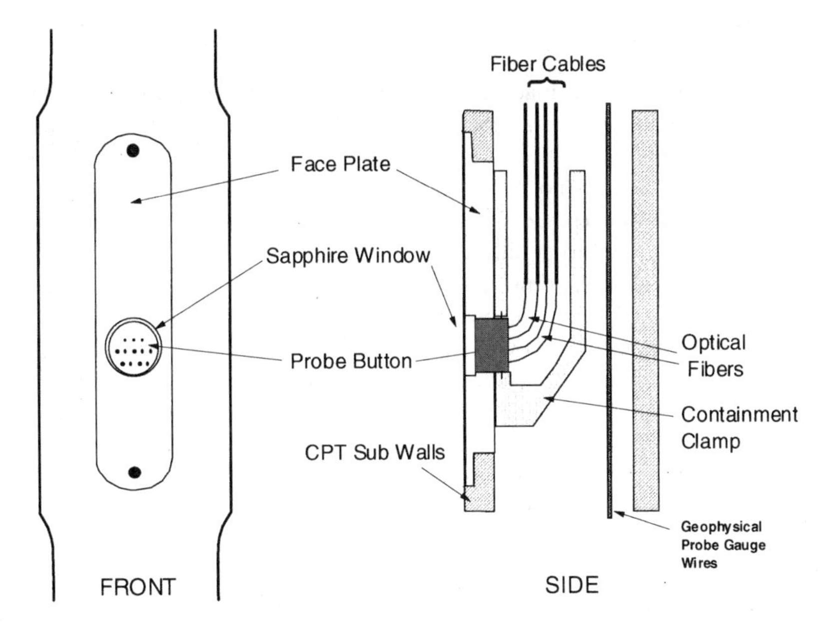

Figure 5 Probe assembly and CPT sub. Individual fiber pairs are seen on the front view of the probe button. Only four out of six fiber cables are shown. Each cable contains four optical fibers.

The optical cable in the LIF system is configured in three segments, one from the laser excitation table to the connection panel, consisting of the 10 fibers carrying excitation light to the sample; another from the probe to the patch panel (threaded inside the CPT rods), consisting of 10 excitation–emission fiber pairs; and a third segment from the patch panel to the detection system, consisting of the 10 fibers carrying emission to the spectrograph (Figure 6). These segments are connected to each other at the patch panel using ST-type connectors and adapters. Connectors for the 400/440/480 μm multimode-type fibers are specially overdrilled because standard connectors accommodate a maximum fiber diameter of only 125 μm. Coupling losses associated with these connectors (1 dB) can be reduced to 0.2 dB using index-matching fluid. Cutoff filters to block laser backscatter and transmit the longer-wavelength fluorescence[14] were inserted into the emission fiber connectors at the patch panel.

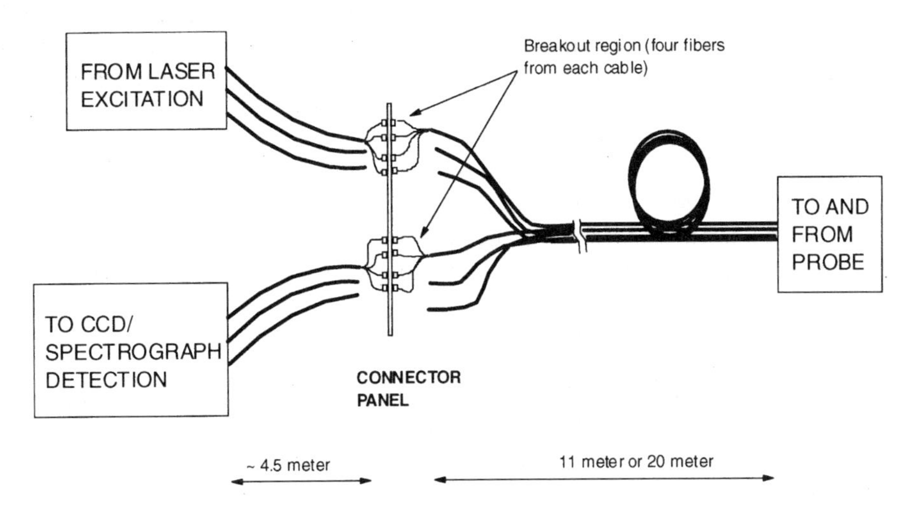

Figure 6 Layout of optical fiber cables used in LIF-EEM–CPT. Breakouts are only specifically presented for two of the connected cable pairs.

2.4.4 LIF Probe

Two significant changes were made in this LIF-EEM probe design from the previous EPA version. First, the orientation of optical fibers at the probe window was altered to minimize direct back-reflection from sapphire window surfaces and sample.[18] This was accomplished by tipping both fibers at off-normal angles (not 90°s) to the window surface. Second, all fiber pairs were mounted on a "button" of diameter less than 1 cm to obviate the need to construct single-depth EEMs from emission spectra taken at different probe depths.[19] In the first-generation multichannel probe, there were 10 individual windows that were each separated by 1.5 in on the vertical axis of the subassembly. There is only one window in the second-generation probe.

The probe fits into a stainless steel CPT subassembly, the "sub," which is attached to the cone penetrometer tool. It includes a transparent sapphire window that is in contact with subsurface soils, a stage that rigidly holds optical fibers in correct orientation against the window surface, and protection for optical fibers inside the sub segment (Figure 5).

The orientation of optical fiber tips has progressed from parallel fibers with flat-cut faces oriented perpendicular to the window surface to fibers oriented at different angles in an effort to improve signal collection efficiency in the probe.[20] Numerical aperture, window composition, and window thickness all dictate the paths of light rays, which affect the overlap between excitation and detection regions of optical fibers in contact with the outer sapphire window surface. The two-fiber arrangement in the current probe maintains overlap of the excitation and detection surfaces and reduces detection of backscattered light between the sapphire window and the soil, which reduces a significant interference in the fluorescence signal.[18] Crucial machining tolerances in this design include the distance and angle between the two fibers of each pair and the angle between the window and the symmetry axis between the fibers.

When using a design with multiple fiber pairs, it must be assured that the excitation/detection regions for each channel are separated from those for the other channels. The presence of overlap in these excitation/detection regions would result in cross talk between the channels, which would corrupt the EEM data. The first-generation probe design (10 windows) prevented the overlap problem by separating the fiber pairs for each wavelength channel by 1.5 in. However, this separation occurred vertically, which resulted in each channel sampling soil at a different depth. Data had to be depth corrected before an EEM analysis could be completed. In the current probe design (one window), the distance between channels is only 2 mm, and no depth correction is needed.

Performance testing has confirmed that cross talk between the fiber pairs is minimal at the 2-mm spacing. Backscattered light at the excitation wavelength remains strong even though it has been significantly reduced compared to the first-generation probe. This could be caused by diffuse scattering from the heterogeneous soil samples, which suggests that performance is in part based on the nature of the soil matrix. This cross talk feature is spectrally narrow and can be readily subtracted from the fluorescence signal. A more severe type of cross talk could occur if fluorescence response from one channel were to be acquired by a neighboring channel at the probe window and soil interface. The LIF-EEM technique relies on completely independent fluorescence information content from channel to channel, and all experimental tests indicated no identifiable cross talk of this type. One of the tests involved placing a white fluorescent card over the probe window and sending a single excitation wavelength through one of the channels. It was observed that, although laser back-reflection peaks were seen in other channels, fluorescence was only seen in the appropriate excitation channel.

A fiber holder, or "button" was designed to provide rigid, angled orientation of the fibers and facilitate close spacing of 12 channels (24 fibers) in one sapphire window (Figure 5). The button consists of a small solid cylinder of aluminum drilled with 24 holes, the diameters of which are slightly larger than the optical fiber (480 μm). Each hole is machined with a small recessed area around it to hold the epoxy used to assemble the fibers. Fiber tips and hardened epoxy are then polished flush with the original button surface.

The button is housed within a three-component assembly that is fixed to the subwall by screws in a faceplate. The button assembly is rigidly fixed inside the probe to enhance the durability of the probe; no springs are used to hold fiber tips against the window, as in other probe designs. Therefore, another crucial machining tolerance in this design is the distance between the inner sapphire window surface and the button.

Another aspect of probe performance that is evaluated using the white card is the collection efficiency of detection fibers for each channel. Relative "overlap factors" for channels in each probe were measured by transmitting a single excitation wavelength (354.7 nm) and measuring the fluorescence response from the white card. Normalized overlap factors obtained with this method ranged between 0.25 and 1.0. Although not every channel exhibits the maximum excitation–emission efficiency as designed, the variations are accounted for using the energy normalization scheme. Overlap factors obtained from the card must be included, along with laser intensities, in any correction procedure that attempts to normalize the channel LIF responses on soil samples to produce a corrected LIF-EEM. Future probe button designs should rigorously define machine tolerances. The relationship between dimension errors and overlap factors is under evaluation.

Overlaps were also measured in solution where greater penetration of excitation light into the sample occurs. The result, as expected, was that white card and solution measurements for a given channel and given probe did not yield equally scaled overlaps. Overlaps in solution depend on optical properties of the solution and on probe geometry. A theoretical model is being developed to predict fluorescence response in solutions for any channel. Overlaps obtained from white card measurements will be used to check the accuracy of the model as they represent the limiting case of no penetration depth.

2.4.5 Detection System

To record multichannel fluorescence, the detection system incorporates an optical fiber interface, an imaging spectrograph, and a charge-coupled device (CCD) detector (Figure 7). The spectrograph and detector systems were obtained as an integrated unit that included Windows environment software (SpectraMax for Windows™). Operation in non-DOS format provided the capacity for multitasking. This spectrographic instrumentation is capable of collecting the full fluorescence signal as the probe is advanced. This signal is manipulated mathematically to form the EEM.

The spectrograph is larger and has better imaging capability than the previous LIF-EEM system to accommodate a larger CCD detector chip and to reduce field curvature distortion from the optical components. Better optical imaging also reduces a source of cross talk between channels that arises from blurring of information along the vertical dimension of the CCD. The larger spectrograph also accommodates two gratings that can be interchanged by software commands. The optical fibers are arranged vertically, in alignment with the spectrograph entrance slits, using an adapter plug. The fiber ends are positioned close to the plane of the slits to improve alignment and to optimize signal throughput on the fiber channels.

The CCD chip has a larger active area (2000 × 800 pixels, each 0.015 mm) than comparable chips to provide better image resolution and less cross talk with greater peak quantum efficiency (65% compared to 45% for standard chips without back thinning). The chip is thermoelectrically cooled and back thinned with UV-enhanced antireflection coating to monitor fluorescence wavelengths between 250 and 1000 nm. The operating software provides sufficient binning to separate the 10 channels of fluorescence data, where each channel can have a variable superpixel width. The software also has a calibration feature that can create files with a calibrated wavelength x-axis.

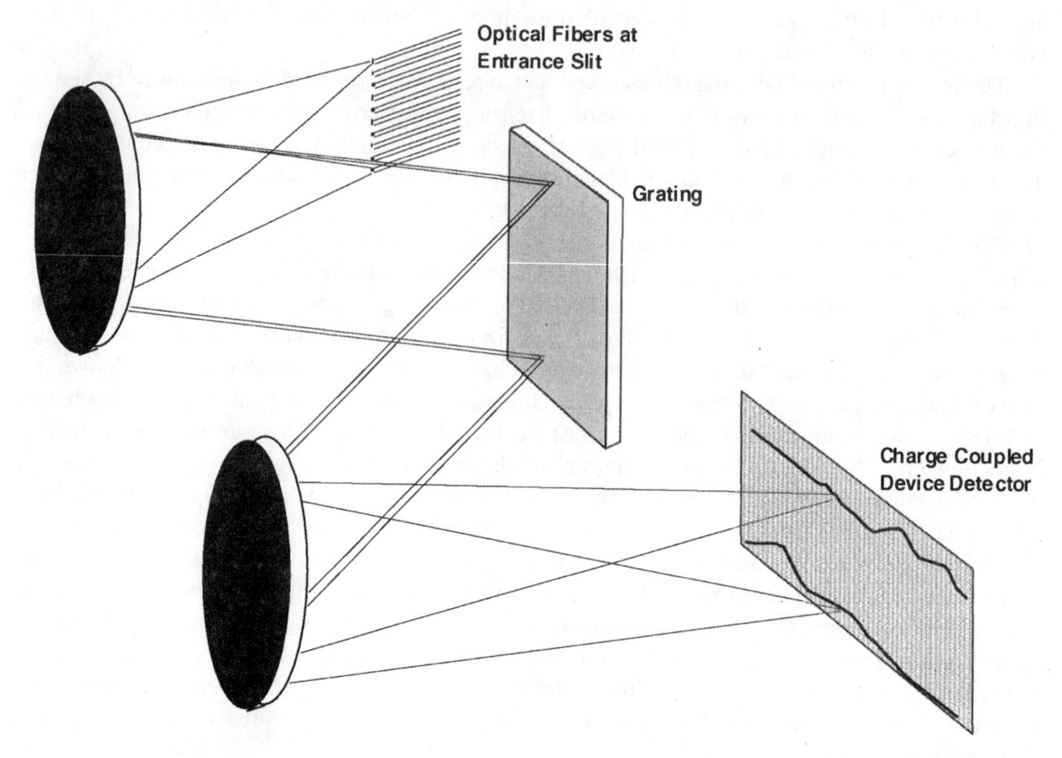

Figure 7 Schematic of multichannel LIF-EEM data acquisition system consisting of spectrograph elements and CCD detector.

2.5 LASER ENERGY CALIBRATION AT THE PROBE

The output energy from the Nd:YAG laser was monitored using a rhodamine B (RhB) solution. RhB was selected as the standard because it not only absorbs strongly at wavelengths above 500 nm but also has low-level absorption in the LIF range of UV wavelengths, effectively absorbing all excitation wavelengths delivered by the LIF probe. Other standard solutions, such as quinine sulfate, did not meet this requirement, and potential solid standards exhibited spatial inhomogeneities. A flat, solid standard would be preferable to a solution because the solid is portable, easily placed at the sapphire window, and requires no preparation.

Two procedures were investigated for calibration over the course of the project. The first procedure utilized two energy meters, one at the laser table with a quartz beamsplitter and the second measuring laser output at the probe. The meter on the laser table monitored pulse energy variation as RhB solution counts were being obtained with the probe. For each wavelength, the ratio of the two measured values was determined. In this way, the laser energy was determined by converting the measured energy at the table to that at the probe, using the measured ratio. In the second approach, the laser output at the probe was measured immediately before and after each RhB EEM was taken, and an averaged energy of five measurements was obtained. The latter approach became the preferred method of calibration because such measurements were simpler and more efficient than the dual-meter approach and it also maintained energy vs. RhB linearity. Two examples of calibration curves obtained in the laboratory are shown in Figure 8, with laser output energy converted to photons for comparison of different excitation wavelengths. The linear correlation coefficient (R^2) is 0.99 for most of the curves.

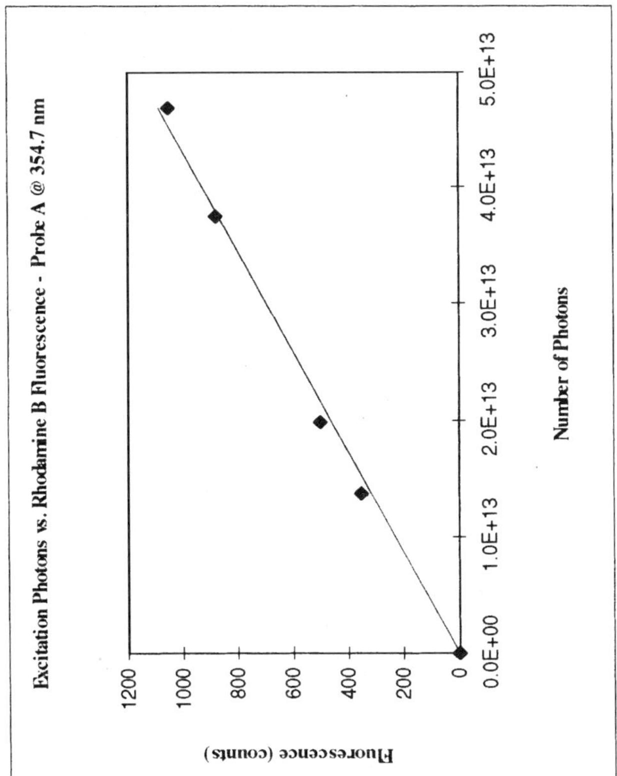

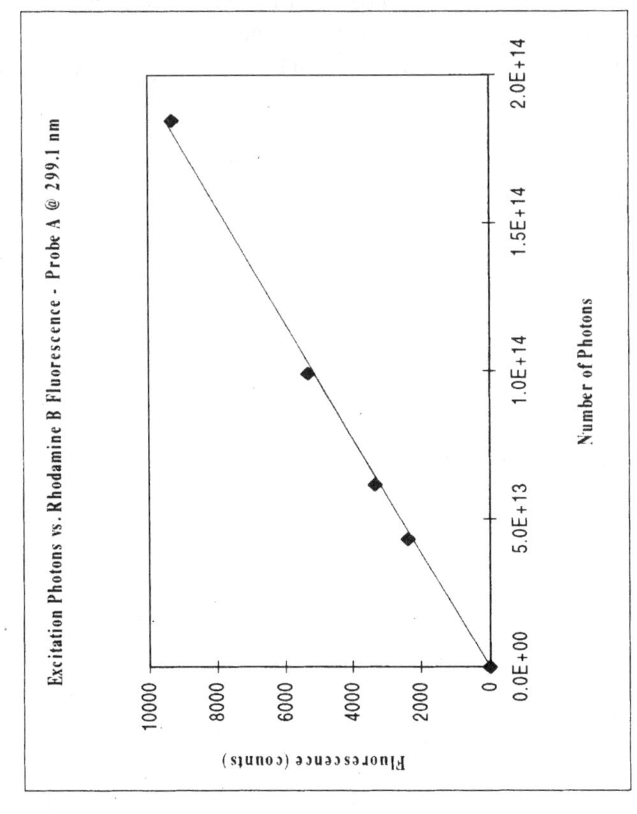

Figure 8 Examples of correlation of excitation photons and rhodamine B fluorescence.

Before and after each push, the probe is dipped into a large cylinder filled with the RhB solution (0.2 mg/l in reagent alcohol) beneath the CPT vehicle. The cylinder diameter is large enough that the cylinder walls do not limit penetration of excitation light into the RhB solution and long enough that it accommodates the length of the probe and CPT tip. The container is placed underneath the CPT vehicle so that the cone tip can be advanced directly into it, using the standard hydraulics without any disassembly. To avoid cross-contamination, the tip is rinsed with ethanol before and after insertion into RhB solution. RhB fluorescence is then measured using the standard procedure for EEM collection so that variations in fluorescence excitation energy can be estimated within the time span of the push. Channel counts are converted to energy using a predetermined probe calibration.

Channel calibration for each probe is performed in the laboratory by determining the linear relationship of RhB fluorescence counts to varying laser energy at each wavelength. Small variations in geometry of each two-fiber pair affects the efficiency with which RhB fluorescence is collected at the probe window. Therefore, calibrations are performed for every channel of every probe, and records are kept of which probe was in use at a given time. Alteration of the probe (e.g., switching a wavelength to another channel on the probe or changing fiber length) requires acquisition of a new energy vs. RhB fluorescence count relationship. As a result, the experimentally obtained linear relationship includes the effects of variations of laser output energy, photophysical properties of the calibrant (absorption at the excitation wavelength, fluorescence efficiency), optical fiber attenuation, overlap of the excitation and detection surfaces at the probe/sample interface, and spectrograph imaging.

Figure 9 presents calibration results (including baselines) obtained for four different probes, with two different optical fiber lengths, for one excitation wavelength (288.4 nm). The close linearity obtained between RhB fluorescence counts and measured laser energy is clearly displayed and noteworthy for probes having two different optical fiber lengths. Another aspect of probe calibration, illustrated in Figure 9, is variation of calibration graph slopes between the probes, arising from machining variation in excitation/detection fiber-pair geometries.

Before the first field test, these calibration graphs were constructed for all wavelengths except those with the weakest energies (246.9, 257.5, and 327.6 nm), which were below the detection limits of the energy meter. Analysis of thc results of the first field test led to development of a signal correction protocol based upon probe channel corrections and spectroscopic properties of RhB, which allowed RhB counts to be converted to energies for the weaker laser output energies. In later calibration measurements, a more sensitive silicon detector was used (instead of pyroelectric) to provide more reliable laser energy vs. RhB counts plots at the weaker wavelengths for data from the second field test.

The original alignment and calibration of the LIF detection system and the initial calibration of the LIF probe are described in the AATDF report "Operation and Calibration Manual for Laser Excitation–Emission Matrix/Cone Penetrometer Tool," which is included in Appendix B.

2.6 EEM DATA ANALYSIS

The EEM data collected using the LIF-CPT probe required pretreatment, which included the subtraction of the baseline, the removal of excitation-scattering light, and the correction for power normalization. Subtraction of the baseline was accomplished by subtracting a background EEM (i.e., an EEM obtained using the same experimental conditions but where no fluorescence signals were detected) from all EEMs. The excitation-scattering light was removed by deleting the narrowband scattering peak from the corresponding laser excitation wavelength and then regenerating the data points by curve fitting. Correction for power normalization was performed using the RhB calibration routine (see Sections 2.5 and 3.2.4). Following pretreatment, the analysis method applied

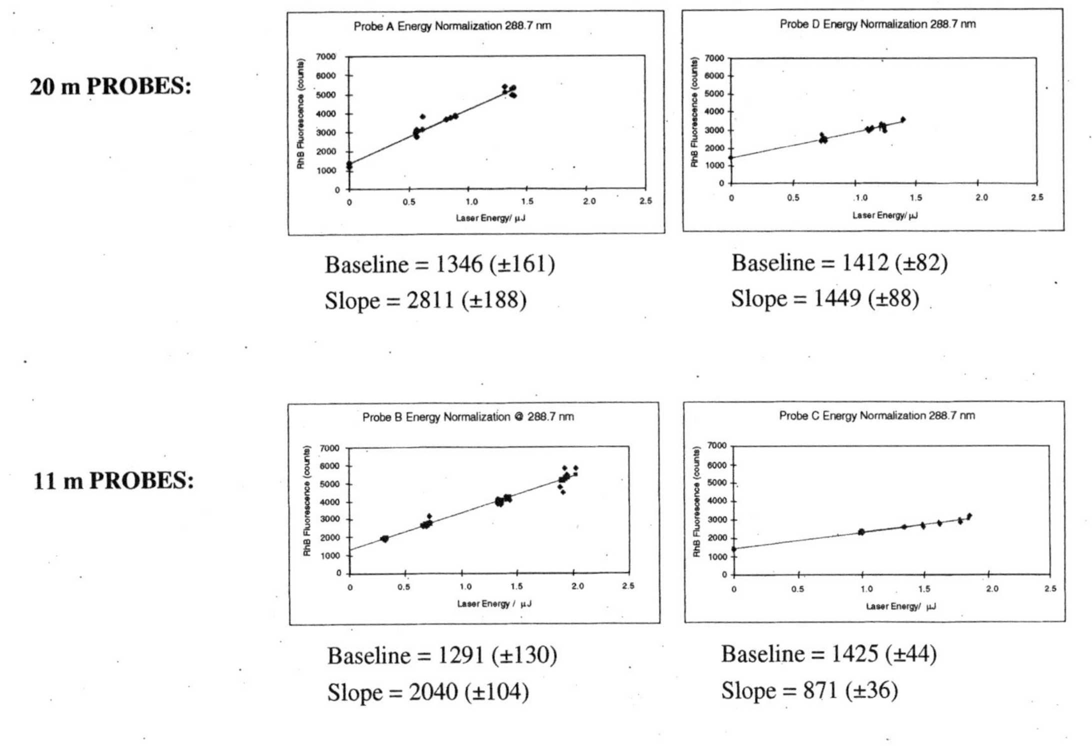

Figure 9 Calibration curves of rhodamine B emission with 288-nm excitation energy for probes having two different optical fiber lengths.

to the field data depended on the quality of the data and the stage of development of the analysis procedures.

During the first field test of LIF-CPT system operations (Hanscom AFB), the data analysis method involved visual inspection of excitation-emission graphs to identify the contaminant and then quantification of the data using summed fluorescence (e.g., for correlation with TPH) or emission at the peak wavelength (e.g., 340 nm for naphthalene), at an appropriate excitation wavelength.

A more rigorous method of data analysis was developed after the first field test, which employed a combination of several factor analysis methods. It was used for the analysis of data obtained during the second field test of the LIF–CPT system, at Otis ANGB. The factor analysis procedure consists of three parts: abstract factor analysis (AFA) to determine the number of fluorescent components in the unknown mixture; target factor analysis (TFA) to identify these components; and rank annihilation factor analysis (RAFA) to quantify the components. Very briefly, RAFA achieves quantitation for a particular test solution by removing a multiple x (i.e., x_A for standard A) of its standard EEM (i.e., $EEM_{std,A}$) from the EEM of the mixture (i.e., $EEM_{mixture}$),

$$\boldsymbol{n}_{\text{LIF, N}} = k\,(N + 2.039M) \qquad (1)$$

to obtain the remainder matrix, R. The factor x_A by which the standard EEM is multiplied is optimized by requiring that the procedure reduce the rank of the mixture EEM by unity, that is, the rank (number of linearly independent rows or columns, whichever is smaller) of R is one less than the rank of $EEM_{mixture}$. This mathematical condition is equivalent to finding the concentration of the test compound in the mixture. RAFA gives meaningful results for a given test compound even when other substances not in the database are contributing fluorescence to the EEM of the mixture, a property that makes it the method of choice for in situ field measurements that preclude chemical separation of different components of a complex mixture.

The algorithms of factor analysis used in this research are based on the works of Malinowski[21] for AFA; Lorber[22] and Malinowski[21] for TFA; and Lorber[23] and Sanchez and Kowalski[24] for RAFA. The algorithms are modified to meet two requirements, which include handling *real* data (rather than computer-simulated or less-than-three component mixture data, as seen in the literature) and operating sufficiently fast to process large amounts of EEM data. The computer programs used in this factor analysis are written in MATLAB™ language and run under MATLAB 5.0™ software. These analyses yield information about the type and the quantity of fluorescent components in the subsurface contaminant.

Factor analysis techniques (AFA, TFA, and RAFA) for EEMs were initially tested by collecting LIF data from a three-component solution of phenol, dibenzofuran, and 2-naphthol. The results, including the number of factors (n = 3, which indicates a three-component mixture), the predicted identity (confirming the presence of phenol, dibenzofuran, and 2-naphthol), and the concentrations of the three components (less than 0.5% error), agreed very well with the known values. The validity of the factor analysis technique was further tested by analyzing the EEMs, obtained by using the LIF-EEM probe, of 11 standard solutions and 9 multicomponent solutions (containing two to five components, compositions given listed in Section 4.3.1).

Finally, this procedure was applied on-site (to core samples) and in situ (during CPT soundings) to EEM data collected during field testing at Otis ANGB. Because the soil core samples were also analyzed by an outside (off-site) laboratory using standard EPA methods, comparison of the field EEM/factor analysis results and the laboratory results was conducted. The comparison is presented in Section 4.3.2.

For easy visualization, Spyglass Transform™ software was used to create the EEM interpolated images from the analyzed fluorescence data. An example of a three-dimensional EEM for a mixture of *p*-xylene, naphthalene, anthracene, and fluoranthene is shown in Figure 10. Examples of two- and three-dimensional EEMs are also included in Appendix C, which shows representative graphs for single component solutions and solutions with mixtures of components measured in the laboratory and for field data (unknown components).

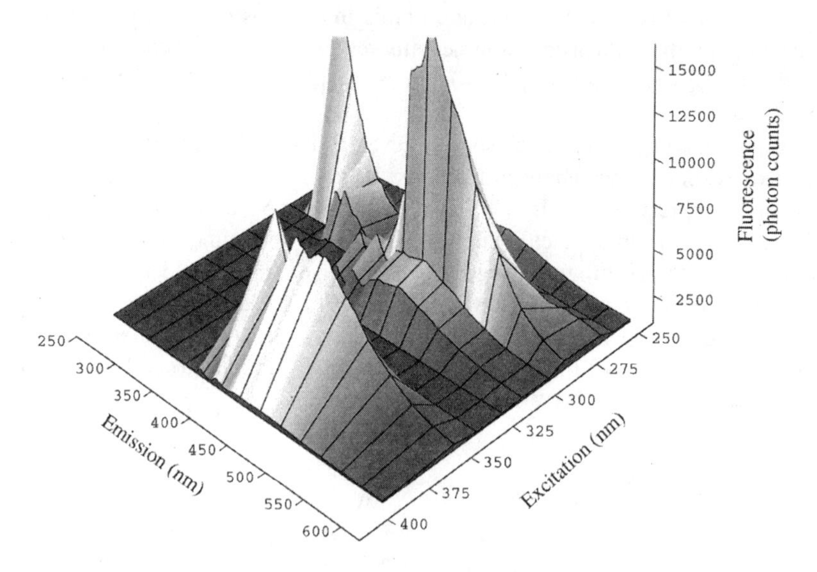

Figure 10 EEM of a *p*-xylene, naphthalene, anthracene, and fluoranthene mixture.

2.7 CONE PENETROMETER TESTING EQUIPMENT

Fugro Geosciences CPT equipment was used for direct push of the LIF probe during the field tests (Figure 11). The LIF probe is deployed through the floor of the CPT truck, which is completely self-contained and climate controlled. The CPT truck houses the operating controls, electrical and optical cables, push rod segments, computers, and ancillary equipment. The vehicle generates AC power for lighting and computer equipment.

Figure 11 Cone penetrometer truck provided by Fugro Geosciences for Hanscom and Otis field tests.

The CPT truck uses a hydraulic ram mounted near the forward portion of the truck to push 1-m-long, threaded steel rods into the ground. These rods were used to advance the LIF probe by appropriate intervals and conveyed the CPT electrical and LIF optical cables from the down-hole instruments to the on-board computers.

During field testing of the LIF probe, the CPT vehicle was accompanied by a support truck. The support truck pulled a trailer containing a high-pressure steam cleaner, grout mixer, and pump. The steam cleaner was used to clean down-hole equipment before each sounding. The grout mixer was used to prepare grout and cement mixtures for sealing CPT sounding holes following data collection.

CHAPTER 3

Summary of Technology Demonstration

Engineering work on the laser-induced fluorescence–excitation-emission matrix (LIF-EEM) probe began at Tufts University where it was designed, fabricated, and tested on laboratory-prepared soil samples and solutions. Following this initial design and development work, the probe was integrated with a cone penetrometer technology (CPT) truck provided by Fugro Geosciences for testing and demonstration in the field.

Two field tests were performed to demonstrate the ability of the LIF-EEM probe to collect semiquantitative data on classes of organic compounds in subsurface soils and to demonstrate advantages of LIF instrumentation that uses simultaneous multiple wavelength excitation.

The first field test was conducted during February 1997 at Hanscom U.S. Air Force Base (AFB), near Bedford, Massachusetts. Goals for this field test included the evaluation of the LIF-EEM equipment array and operational parameters, development of equipment calibration techniques, comparing and contrasting the results from in situ fluorescence measurements and those from laboratory analysis of soil samples, and providing an understanding of the nature and extent of contamination at the test site. Data obtained during the first field test were used to modify the instrumentation, software, and operating procedures to optimize and streamline performance.

The second field test was then performed using improved procedures. The second test was completed during August 1997 at Otis U.S. Air National Guard Base (ANGB) on Cape Cod, Massachusetts. This test provided demonstration of the LIF-EEM technique. Qualitative and semiquantitative information was obtained on subsurface soil contaminants, and characterization of a small contaminated area was completed. The results, in turn, were used to further fine tune the instrument.

3.1 GENERAL SCOPE OF FIELD WORK

Prior to each field test, the LIF equipment was integrated with the CPT truck at the Tufts University campus. The equipment was then mobilized to the site for testing. The probe and instrumentation were housed in the CPT truck, which included ancillary equipment for performing cone penetrometer soundings, generation of required electrical power in the field, probe decontamination, and grouting of test holes.

Field work at each test site involved advancement of the LIF probe through soils by the CPT equipment and collection of fluorescence data using the LIF-EEM instrumentation. Some of the positive fluorescence results were evaluated for validity by soil sampling within several feet of the original test hole. The collected soil cores were then analyzed for selected contaminants using U.S.

Environmental Protection Agency (EPA) analytical techniques to confirm the presence and concentrations of contaminants. At Hanscom AFB, the LIF probe and a TDGC/MS probe developed by Professor Albert Robbat, Jr., also from Tufts University, were passed over selected soil cores to compare data from the two instruments.

3.2 FIELD TEST NO. 1: HANSCOM AIR FORCE BASE

Hanscom AFB is located at the intersection of the towns of Lincoln, Lexington, and Bedford in Massachusetts. Field test No. 1 was conducted at IRP Site 21, which is a former fuel-storage tank farm (Figure 12). Fuel management facilities in the IRP Site 21 area included off-loading, storage, and dispensing facilities for jet fuel, aviation gasoline, and, more recently, No. 2 fuel oil. The source of the petroleum contamination at the site is believed to be the former above-ground and underground jet-fuel storage tanks (ASTs and USTs, respectively) and associated piping.

IRP Site 21 storage facilities included two 500,000-gal above-ground jet fuel storage tanks, five above-ground 50,000-gal aviation-gasoline (AVgas) storage tanks, and six underground 12,000-gal jet fuel/AVgas storage tanks. The site also contained fuel loading/off-loading stands, a rail siding, and pump houses. During the 1970s, the westernmost 500,000-gal tank was used to store No. 2 fuel oil. The aviation system was taken out of service in 1973, and the 50,000-gal tanks were removed in 1986. The 500,000-gal tanks were abandoned in 1973 but not removed until 1990 when they were pumped, cleaned, and dismantled.

3.2.1 Site Conditions

Soil beneath IRP Site 21 is typically comprised of sand and silty sand sediments from a depth of about 0 to 15 ft, underlain by silty clay beds and an irregular layer of glacial till. The glacial till generally prevented advancement of the CPT soundings below depths of approximately 15 to 18 ft. Groundwater was typically encountered below a depth of 12 ft, as monitored by existing wells.

During previous investigations, soil samples were collected from a limited area near the former underground fuel pipeline and chemically analyzed using U.S. EPA methods. Results indicated elevated concentrations of BTEX, naphthalene, and total petroleum hydrocarbon (TPH) ranging up to 18,000 ppm. Data from previous EPA LIF-CPT investigations and product recovery from wells also identified the presence of separate phase hydrocarbons in soils at a depth of about 6 to 12 ft.

3.2.2 Field Operations

Field test No. 1 was conducted from February 18 to March 7, 1997. Equipment installation in the CPT vehicle and associated tasks were completed within 4 days. A total of 26 LIF-CPT soundings and five soil-sampling CPT soundings were completed over an 11-day period. The locations of the soundings are shown in Figure 13. Fourteen LIF-CPT soundings and all soil sampling were conducted near Building 1843 where No. 2 fuel oil was detected in ground water recovery wells. The other LIF-CPT soundings were conducted near the former 500,000-gal storage tanks, at Building 1823.

The LIF instrumentation was powered by an internal generator on the Fugro CPT truck. The generator and instrumentation were turned off each time the CPT vehicle moved to the next sounding location, and instrument warm-up time after being turned on again was approximately 30 to 40 minutes. During a push, the CPT advancement was stopped at increments of 2 to 6 cm to acquire a LIF-EEM spectrum. The length of time for a single push ranged from 45 minutes (no emission detected) to about 2 hours (strong emission signal detected over a wide range of depths).

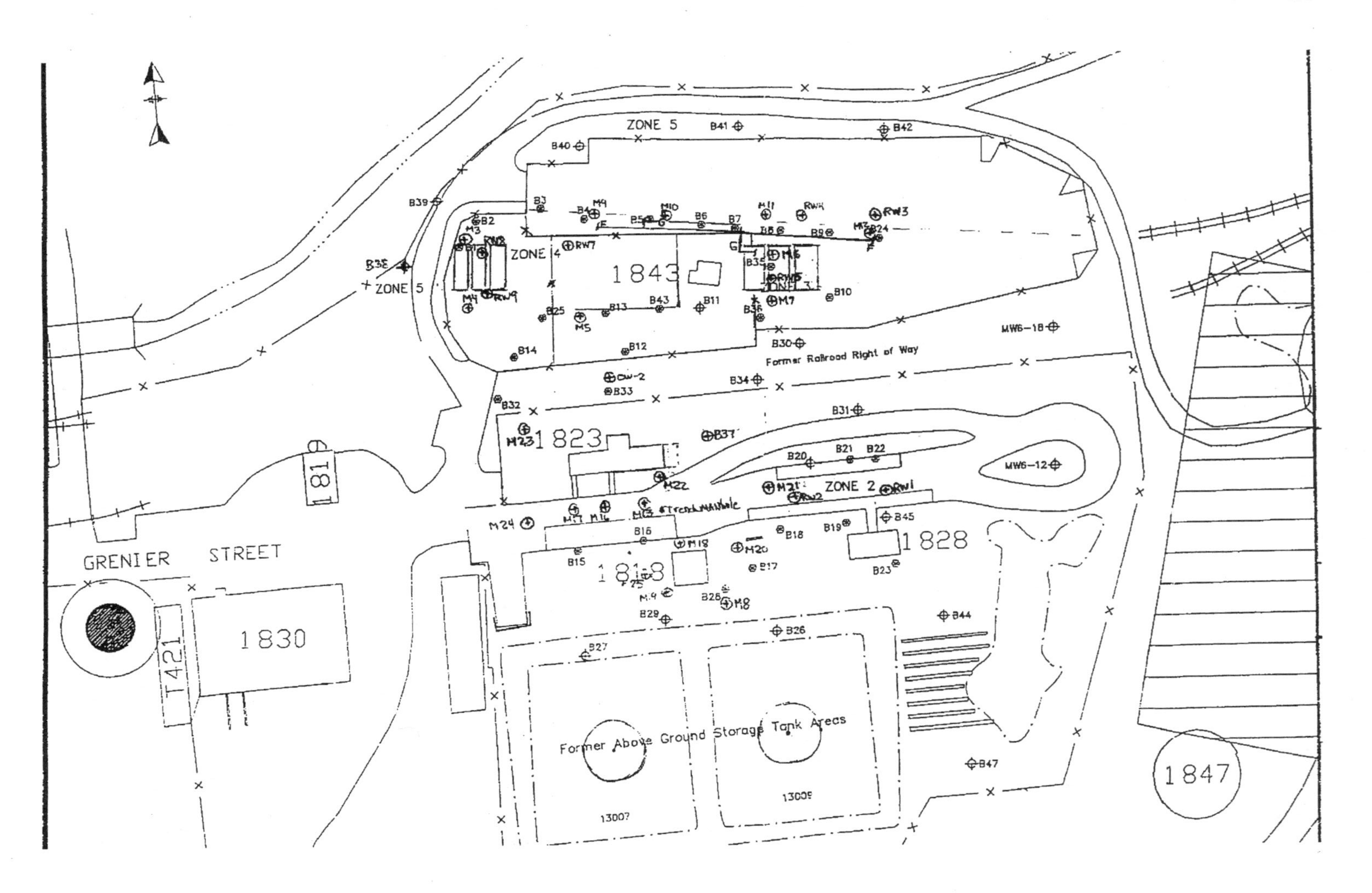

Figure 12 Map of Hanscom Air Force Base, IRP Site 21.

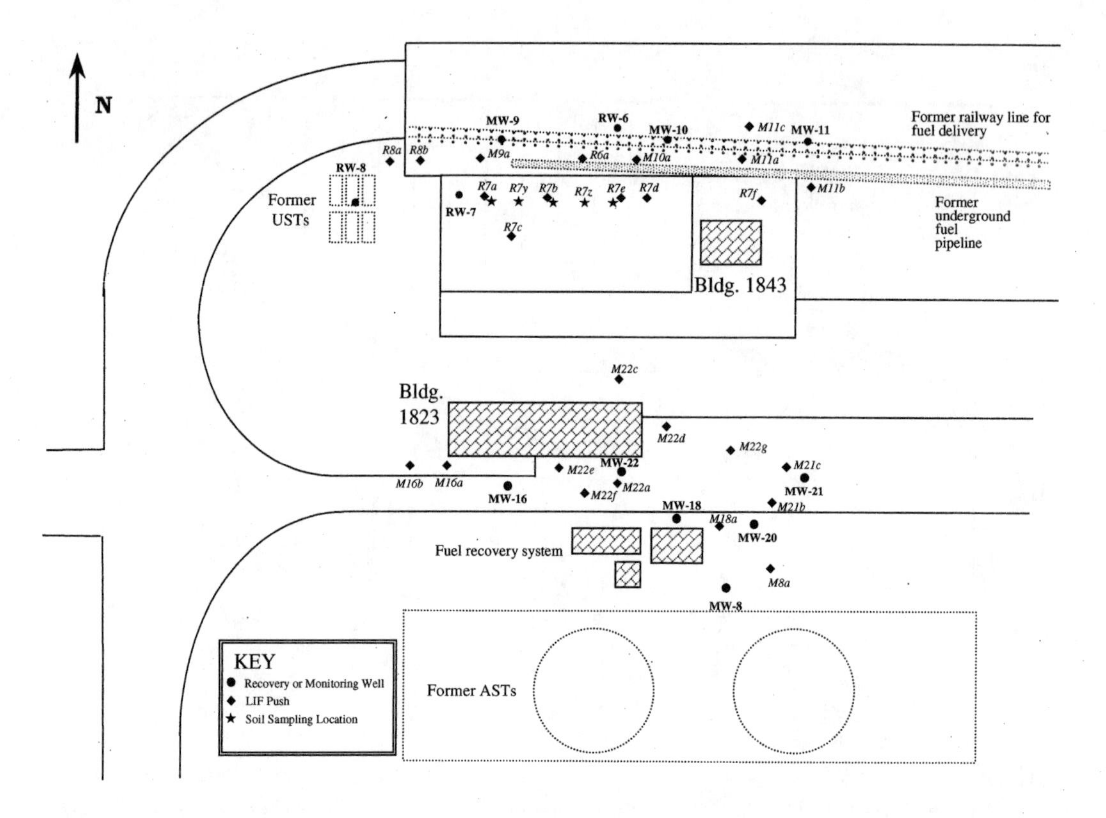

Figure 13 Hanscom Air Force Base site map with LIF-CPT push locations.

3.2.3 Soil Sample Collection and Analysis

A component of the field activities included soil sampling and analysis to validate the LIF-EEM data. On-site LIF probe measurements, on-site gas chromatography/mass spectrometry (GC/MS) analyses, and off-site EPA-method laboratory analyses were all conducted on common soil cores recovered with the CPT. The site map (Figure 13) shows the locations of soil samples collected near the former underground fuel pipeline. Depths of the soil samples collected at each location were 100 to 130, 215 to 245, 260 to 290, 305 to 335, 360 to 390, and 430 to 460 cm. A total of 29 cores were collected for soil sampling and field analyses.

Soil cores were obtained using a CPT soil sampler consisting of a 1.5-in diameter by 36-in length stainless steel core barrel fitted with a conical tip. The tip was equipped with an o-ring seal to prevent water and soil from entering the sampler during advancement to the target sampling depth. At the desired depth, the conical tip was retracted upward through the sampling tube by a release mechanism from inside the vehicle, and the tube was advanced approximately 36 in for collection of the soil core.

Soil from the core was mixed to form a composite sample, and a small portion of the composite sample was subjected to on-site analysis by an LIF-CPT probe (not attached to the rod system), followed by on-site GC/MS analyses performed by the Robbat research group at Tufts. The remainder of the samples were analyzed off-site by an environmental laboratory (Groundwater Analytical Laboratory, Buzzards Bay, MA) using EPA methods. In this way, contaminant data was collected and compared using three analytical methods.

On-site LIF measurements with the probe were conducted by placing the soil in a formed metal boat and clamping the probe onto the soil to simulate subsurface pressures. These same samples were sealed into plastic bags and refrigerated in the mobile laboratory for on-site GC/MS measurements. Soil samples submitted to Groundwater Analytical Laboratory were analyzed by the following U.S. EPA methods:

- 418.1 (Modified) for total petroleum hydrocarbons (TPH)
- 8270/PAH for semivolatile organics
- 8100 (API Modified) for diesel range organics (DRO, C_{10}–C_{28})
- 8260 (API Modified) for gasoline range organics
- 8260 for volatile organics, including chlorinated solvents and BTEX

3.2.4 Equipment Calibration

A standard rhodamine B (RhB) solution (0.2 mg/l solution in reagent alcohol) was used to collect fluorescence measurements at the probe window for instrument calibration and normalization of fluorescence data (see Section 2.5). Prior to the field test, RhB calibrations were performed on all LIF channels in the laboratory using the same instrumentation settings, such as slit width and integration time, as for field LIF soundings. Linear relationships were obtained between fluorescence signals and excitation channels at 266, 289, 299, 315, 342, 355, and 416 nm. Energy levels were at nJ levels for the remaining channels (247, 258, and 327 nm), whereas the energy probe had a sensitivity of 0.01 μJ, so linear relationships could not be obtained. These channels were normalized to the intensities of RhB fluorescence, which indirectly indicated excitation energy levels. A silicon energy probe with a sensitivity of 0.01 nJ was procured for the second field test.

In the field, five RhB fluorescence measurements were averaged for each excitation channel before and after each sounding. For the strong channels, the average of the pre- and postsounding RhB fluorescence signals was converted to the absolute number (n) of excitation photons at the sample, using the fluorescence-energy calibration and the relation $E_{photon} = nhc/\lambda$, where E_{photon} is the energy of excitation photons, h is Planck's constant, c is the speed of light, and λ is the excitation wavelength. It was observed that the 299-nm excitation channel, which is capable of exciting a wide range of aromatic compounds, yielded the highest signal/noise ratio. After the field test, RhB fluorescence data were measured again in the laboratory and compared to pretest data to detect any degradation in the RhB stock solution used in the field. The data comparison showed a difference of only 3% and a standard deviation of 5%.

3.2.5 Summary of Equipment and Software Performance

Field test No. 1 was the first opportunity to evaluate the LIF-EEM equipment and operational parameters in the field and to develop equipment field-calibration protocols. In general, the LIF instrumentation performed well throughout the field operations. The Nd:YAG laser operated well in the field using the CPT truck internal generator, although periodic down time (about 30 minutes) was required when moving the CPT vehicle. The down time was later eliminated, when an external power generator enabled the laser to operate continuously (see Section 5.2). The optical path alignment, including the Raman shifter and other optics on the laser breadboard, was easily performed in 5 to 10 minutes. The fiber probe design withstood the pressure of the CPT operation, and no optical fiber or probe replacement was necessary during the 26-push Hanscom field test. The power normalization method, i.e., measuring the RhB energy response of each channel before and after each LIF-CPT push, also worked well. With a portable heater in the CPT truck, the cold weather during the field work did not show much effect in the field operation. A more detailed analysis of the performance of the system is presented subsequently.

3.2.5.1 Laser System Performance

The laser output energy showed reasonable stability during LIF-CPT experiments conducted during a single day, with fluctuations as small as 5%, for channels 266, 299, and 355 nm, and fluctuations generally less than 30% for the remaining seven channels. The stability from beginning to end of a push was considerably better, with fluctuations ranging from lows of 1.6 and 1.8% on channels 299 and 355 nm, respectively, to highs of 25.3 and 23.0% on channels 258 and 342 nm, respectively. This suggests that the optical table was reasonably well isolated from truck vibrations. However, large fluctuations over the course of the 9 days of field work were observed for some excitation beams. Possible reasons for this instability may include flashlamp aging, competition between third and fourth harmonic generation, and insufficient temperature control.

3.2.5.2 Probe Performance

In some cases, backscattered laser light from other excitation channels (the strong 299.1 and 354.7-nm lines) appeared in the emission spectrum of another channel whose cutoff filter had not been designed to block them. The resulting narrowbanded spectral features from backscattered laser light were easily subtracted from the broadband fluorescence data from each channel in the LIF-EEM. This procedure provided satisfactory results except in the BTEX region where weaker fluorescence was expected compared to fluorescence from naphthalene and higher-order PAHs. The presence of stray excitation light at 288.4 and 299.1 nm precluded unambiguous identification of BTEX contamination at the site.

3.2.5.3 Spectrograph Performance

The charge-coupled device (CCD) is divided into 10 distinct horizontal "stripes," each displaying the spectrum of light delivered by a single collection fiber, dispersed along the horizontal axis. Choice of boundaries for each stripe is based on a CCD image of the vertical array of fibers when each is transmitting the Hg lamp line spectrum. Any convenient light source could be used for this procedure; however, Hg was chosen because it also serves as a wavelength standard for calibrating the horizontal axis, with several intense lines spaced across the spectral range of interest. Adjustments were made to fully resolve the image that showed the fiber outputs on the CCD. After optimizing the resolution, detector software was used to define pixel ranges for each fiber channel. The width of each stripe was eight pixels, and each stripe was separated from its upper and lower neighbors by a distance of 50 pixels. Those pixels contained overlapping information from adjacent channels and were excluded from the CCD readout. Resulting multichannel spectra were free of detector cross talk from image blurring.

In the horizontal direction, eight adjacent pixels were binned without compromising spectral resolution to improve signal-to-noise ratios and keep the data file size more manageable. Wavelength calibration was also conducted using a Hg lamp line spectrum. The CCD/spectrograph operating software contained an internal calibration routine that assigned linearized wavelength values to the x-dimension superpixels, based on Hg line assignments. This calibration was accurate to within 2 nm, with best agreement at the center of CCD pixel range.

During laboratory diagnostics and experiments, the detector background measured from dark counts was fairly uniform across the wavelength axis of the CCD for each channel. For 0.4-second integration time, the count level was approximately 1200. Throughout the first field operation, the background counts were not uniform, and they showed structure. The fluctuating background of the fluorescence spectra caused difficulty in developing automated routines for subtracting background counts from actual signal counts. When fluorescence response was low, baseline structure was often comparable in magnitude to fluorescence structure. From subsequent discussions with the manufacturer, it was determined that the probable source of the baseline structure was electric charge from radiofrequency noise associated with high voltage sources (e.g., laser power supply), aggravated by insufficient grounding. During the second field test, a grounding cable was attached to the CCD detector cover, which eliminated the baseline noise. Temperature variations in the CPT vehicle also lead to problems in cooling the CCD, contributing to nonuniform thermal noise.

During the first field test, the spectrograph was equipped with a 200 groove/mm grating used in second order (of diffraction) so that fluorescence and scattered excitation light from 265 to 550 nm would be observed at nominal spectrograph, wavelengths of 530 to 1100 nm. This grating was chosen so that alternative spectra taken with 532-nm excitation, the second harmonic of the YAG laser, could be observed without moving the grating. The 532-nm excitation strategy was never deployed, however, and the choice of grating proved unfortunate because some of the field data appeared to be complicated by overlap of long-wavelength emission (possibly from larger PAH) in first-order and short-wavelength emission (possibly from BTEX and similar small aromatics) in third order along with the intended second-order spectra. These artifacts limited the quantitative interpretability of the full EEMs and dictated the use of simpler alternative data reduction strategies (see Section 2.6). A 300 groove/mm grating blazed at 250 nm and used in first order was exchanged with the original grating for the second field demonstration.

3.2.5.4 Data Acquisition and Analysis Routines

With commercially available software (SpectraMax, ISA), three-dimensional spectra of the subsurface environment were measured when cone advancement was stopped, to acquire data, and was then displayed in real time. Because the LIF-EEM instrument was yet not fully interfaced to the cone penetrometer system, data sets were only saved when the probe was advanced by a predetermined interval and stopped at a known depth below ground surface (bgs), which was also recorded with the EEM. Although data acquisition was not triggered by the depth gauge as designed, it was fairly rapid. A full EEM was obtained in a 10-second interval, which was reduced to 2 seconds in the second field test. The raw data obtained in the field were transferred to Excel spreadsheets, where initial processing, including scattered light and baseline subtraction and photon normalization, was done in a streamlined fashion. The availability of the real-time display, both during probe advancement and static data collection, provided operators with an immediate quantitative measure of fluorescence as a function of both excitation and emission wavelengths and allowed real-time adjustment of the sampling plan (vertical intervals between saved data sets and surface locations of subsequent soundings). The availability of this information for immediate incorporation into the sampling strategy represents realization of one of the most important potential advantages of in situ sampling, which is incorporation of so-called "dynamic sampling plans" for faster, cheaper, and more detailed site characterization.

Complete integration of the LIF-EEM and CPT control systems is a realistic goal. Automatic triggering will provide additional important advantages to the LIF-EEM tool. Those advantages would include collection of automatic depth-calibrated spectra from triggering by pulse-train output of the CPT depth gauge, reduction of push time by collecting an uninterrupted LIF-CPT sounding with no stops to acquire data, and continuous data acquisition with the probe in motion to improve depth calibration and average out small-scale inhomogeneities in soil texture.

Automated external triggering is a realizable objective for the LIF-EEM tool. Two avenues pursued during the project were software development from the spectrograph/detector manufacturer and release of a software driver to facilitate outside development of software using LabView™. An additional possibility is adoption of the system developed by Dakota Technologies to solve the same problem in the ROST LIF-CPT vehicles.

3.2.5.5 System Performance

Table 3 shows laser output energy measured for the different wavelengths and probes with different optical-cable lengths during the project.

Table 3 Measured Energies of Excitation Wavelengths from LIF-CPT Probes

Wavelength (nm)	Probe A 11-m cable (μJ)	Probe B 20-m cable (μJ)	Probe C 11-m cable (μJ)	Probe D 20-m cable (μJ)
266.0	12.0	3.2	1.7	1.0
288.4	1.1	1.8	1.6	1.2
299.1	6.5	3.3	13.9	5.3
314.8	0.13	0.26	.42	0.12
341.5	.18	.66	0.44	0.49
354.7	9.0	7.4	5.6	3.6
416.0	2.7	6.9	4.3	1.8

These energies were measured in the laboratory shortly before the first field test, and they represent typical values found throughout laboratory and field measurements. Varying soil textures, wavelength responses, and wavelength efficiencies for different PAH compounds made it difficult to estimate the minimum energies required for signal fluorescence. However, the excitation energies delivered by this LIF tool were found to be sufficient for detecting PAH concentrations as low as 16 ppm in sandy soil.

Probes A, B, and D were field tested and detected PAH plumes at low and high concentrations (ppms). In the laboratory, probe B was used to detect various levels of fluoranthene (16, 32.5, and 77 ppm, each with 7.5% moisture content). In loosely packed sandy soil, 16 ppm was easily detected by wavelengths at 266.0, 288.4, 299.1, 314.8, and (to a lesser extent) 341.5 nm. However, in packed sandy soil, 32.5 ppm was detected with lower fluorescence at all wavelengths. The CCD integration time was established at 0.4 second, and the slits were set at 0.3 mm.

Figure 14 depicts an example of the EEMs obtained in the laboratory from a solution of *p*-xylene, naphthalene, anthracene, and fluoranthene in cyclohexane. The concentrations of the four components were 580, 10, 2.2, and 1.7 ppm, respectively. The fluorescence fingerprints for these four different components, with one to four aromatic rings, are clearly shown.

3.2.6 Presentation of Fluorescence Data

Fluorescence data were obtained at locations and depths where contaminant-saturated soils were suspected to occur near Building 1843 and the former underground fuel pipeline. A simplified presentation of the multichannel LIF data is provided by graphing total fluorescence vs. depth (Figure 15). These graphs indicate the presence or absence of fluorescent signal in each channel, and they were used to identify the most contaminated zones at the site and the relative contaminant concentrations. They were obtained by adding the fluorescence responses at all emission wavelengths for each excitation channel (summed fluorescence) and normalizing the response. The error bars on the graphs provide an assessment of uncertainties in the total fluorescence data.

Evaluation of the fluorescence data involved comparison to results of conventional laboratory analysis of soil core samples using established U.S. EPA methods. The comparison of LIF results and laboratory results was conducted for naphthalene, total petroleum hydrocarbons (TPHs), and diesel range organics (DRO). Using the linear correlation of laboratory concentrations and fluorescence counts on the core samples, the LIF signals (see Section 2.6 for details) were converted to concentrations (e.g., ppm). These results were applied to the in situ LIF measurements to characterize the Hanscom site.

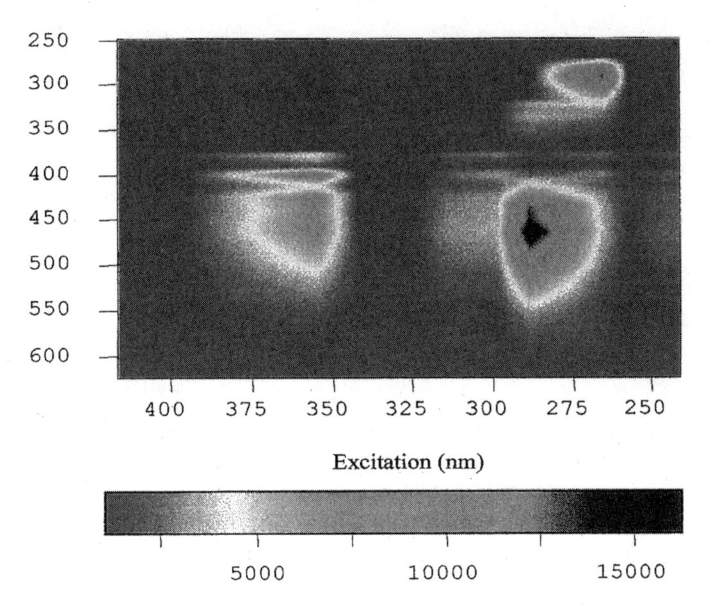

Figure 14 Two-dimensional visualization of EEM of a *p*-xylene, naphthalene, anthracene, and fluoranthene mixture.

EEMs were also constructed from the LIF data, where contamination was detected, and visually compared to the standard EEMs obtained in the laboratory for identification of the contaminants. Results of the evaluations are presented in Section 4.2.

3.3 FIELD TEST NO. 2: OTIS AIR NATIONAL GUARD BASE

Otis Air National Guard Base (OANGB) is located on the Massachusetts Military Reservation (MMR), Cape Cod, near Falmouth, Massachusetts. Historic releases of fuels and solvents have been extensively characterized. Based on these data, five prospective field-test locations were selected for additional site characterization, which was conducted in February 1996 by Fugro Geosciences (Figure 16). No contamination was detected in soil samples from four of the areas (SD-5, PFSA, FTA-3, and Motor Pool near I-Gate). However, soil samples from the Coal Yard-4 (CY-4) area contained appreciable amounts of fluoranthene, pyrene, benzo(a)anthracene, chrysene, benzo(b)fluoranthene, benzo(k)fluoranthene, and benzo(a)pyrene. This site was selected for the second LIF-EEM field test.

The CY-4 area is located near the former Central Heating Plant, adjacent to the East Substation and Building 164, extending south over a 3-acre area (Figure 16). An abandoned rail line trends north–south across the site. Previous site characterizations detected contaminants at the north end, near the former Central Heating Plant where coal used to be stockpiled, and rainfall transported contaminants to adjacent drainage ditch sediments and groundwater. Additional contaminant sources were also suspected in the area of the former Central Heating Plant related to former industrial activities there.

3.3.1 Site Conditions

Soils underlying the CY-4 site are generally silty sand and sand, with interbedded lenses of clay or silty clay at depths above 10 ft. Cobbles were reported in previous drilling at depths below 18 ft.

Previous site characterization at CY-4 detected semivolatile PAHs in soils at concentrations ranging from 0.4 to 22.0 mg/kg. Low levels of BTEX were also reported. Laboratory analyses of

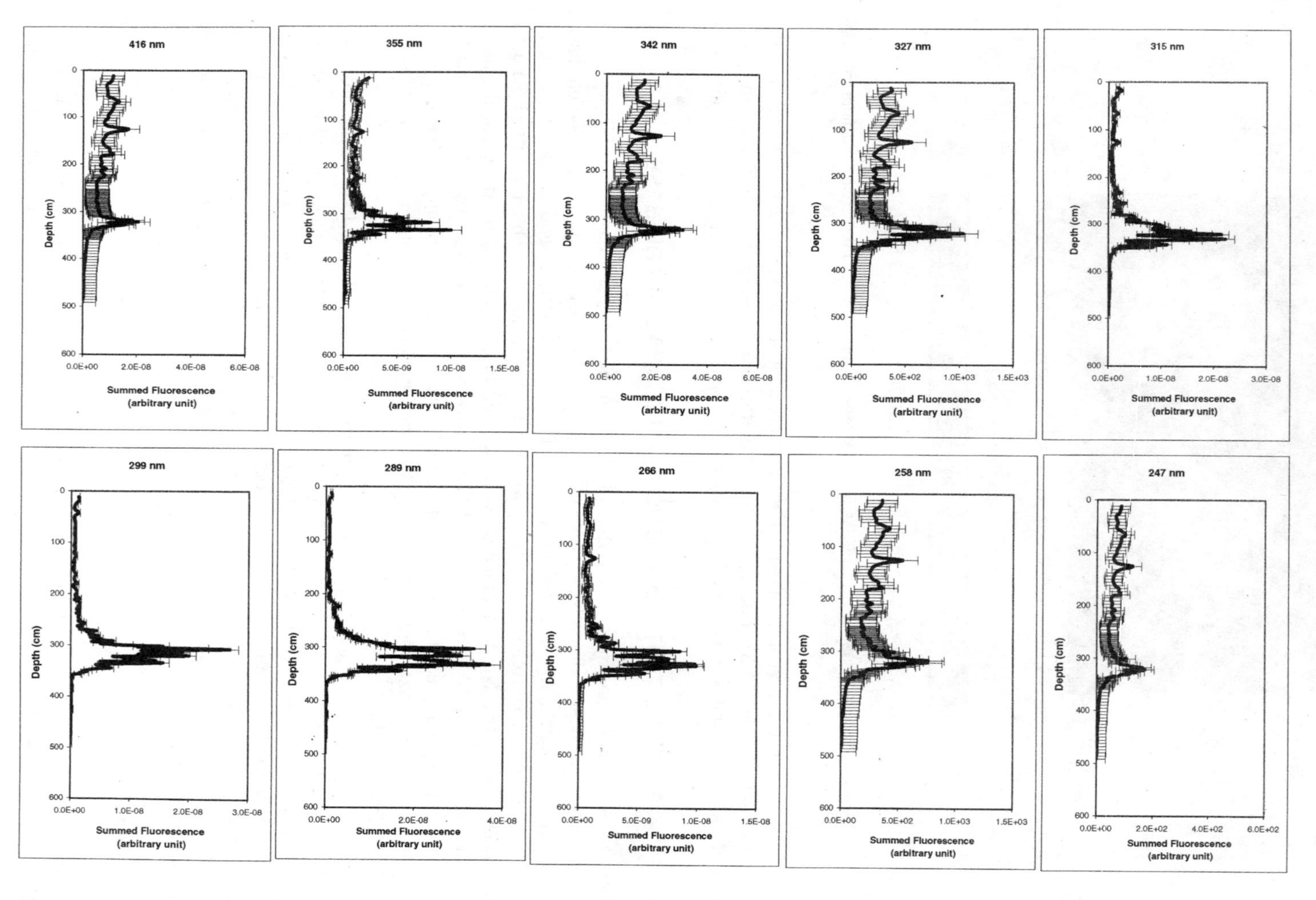

Figure 15 Summed fluorescence obtained at Hanscom AFB site RW-7A.

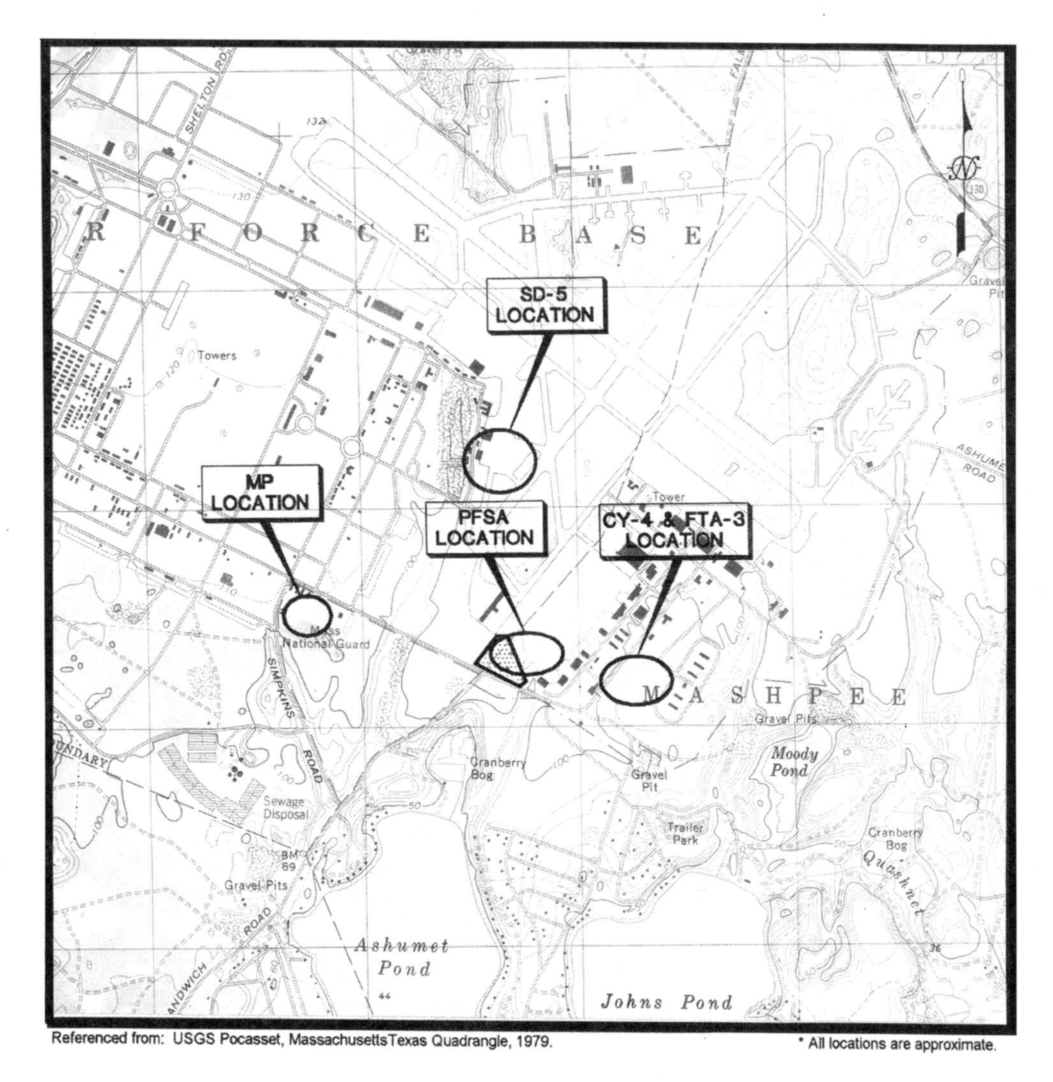

Figure 16 Otis ANGB, Falmouth, MA, locations of additional site characterization.

soil samples collected during this field test indicated concentrations of naphthalene/2-methylnaphthalene up to 63 mg/kg and about 1 mg/kg of heavier PAHs including anthracene and phenanthrene, among others. Concentrations of TPH ranged up to 3200mg/kg, and GC/FID analyses detected kerosene.

3.3.2 Field Operations

Field test No. 2 was conducted from July 23 to August 2, 1997. LIF equipment was integrated with the CPT vehicle at Tufts in less than 2 days, and travel to Otis ANGB took half a day. No extensive troubleshooting was required during this second installation procedure.

A total of 18 LIF-CPT soundings were conducted in the CY-4 area (Figure 17) over a 6-day span. Soils were sampled at two locations. Significant levels of contamination were identified using the LIF data within a localized area of 100 ft^2, with major contributions from naphthalene- and fluoranthene-like profiles.

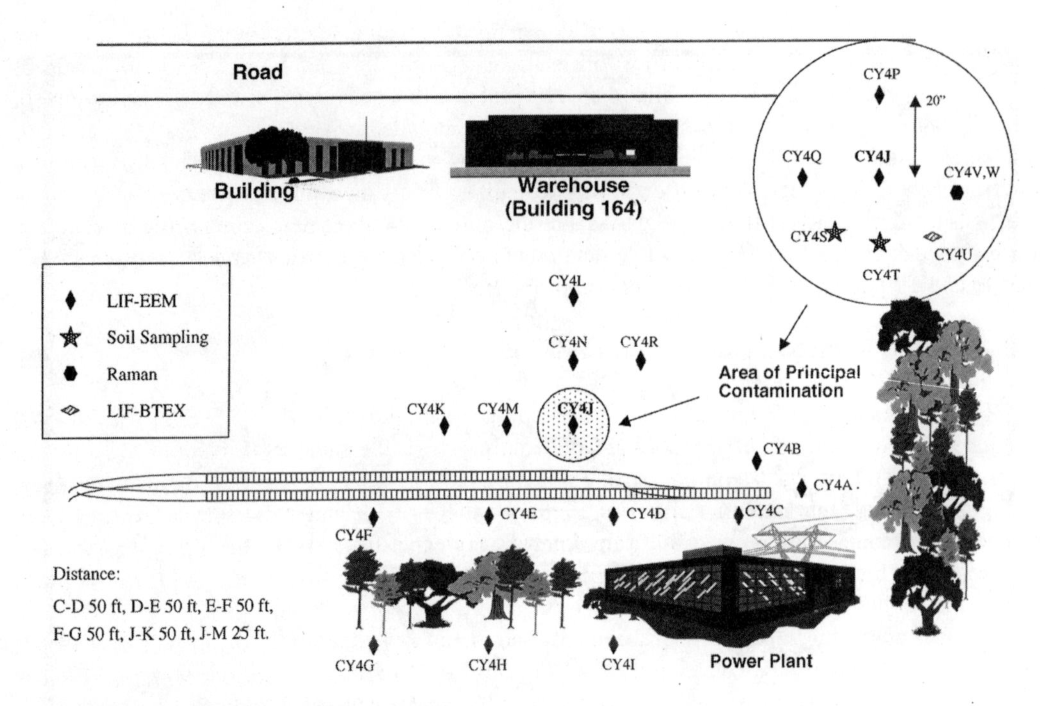

Figure 17 Push locations at Otis Air National Guard Base Coal Yard-4 site.

3.3.3 Equipment Calibration

Rhodamine B (RhB, 0.2 mg/L) solution in reagent alcohol was used as a standard for energy calibration and power normalization. The procedure was similar to that used in the Hanscom field test (see Section 3.2.4). Prior to the field test, the response of each channel of the probe when immersed in the standard solution was calibrated against laser energy at the probe. In the field, the probe was immersed in the standard solution before and after every CPT sounding, and the response was measured to allow correction for variations in the excitation energy for different pushes. The fluorescence signals of RhB measurements for all standards and unknown samples were scaled to the same level, and the scaling factors were used to normalize the sample fluorescence for individual excitation channels.

Two methods were used to obtain calibration curves from these data for different LIF probes. For Probe B, three significantly different energy levels were chosen, the fluorescence signal and simultaneous laser energy were recorded, and this procedure was repeated three times. CCD response was recorded with excitation beam blocked (excitation energy = zero), and data were converted to photons. Ten data points (three energy levels and three points at each level and the origin) were plotted to obtain the photon-calibration curve for an excitation channel. The linear fits for these data generally have R^2 values of 0.95 or better.

After calibrating Probe B, averaged data from five measurements were used, instead of single measurements, to compare output energy vs. corresponding fluorescence. Data from both calibration methods were compared for the 266-nm wavelength on Probe C. The difference in calibration slopes was only 5% (standard deviation for the curves in the first method was 7%), and the linear fit using the second method was better ($R^2 = 0.99$ vs. 0.95). The second method of calibration took half the time to perform, and was used for calibrations of the remaining probes (A, C, and D). R^2 for those probes equaled 0.99 or better. At Otis, Probe A was used for in situ LIF measurements, and Probe B was used for on-site LIF/soil sampling of cores.

In addition to obtaining the fluorescence-energy correlation for the seven strong excitation channels used in the Hanscom field test, measurements were obtained for the excitation energies for the weak channels by using a silicon energy probe with a sensitivity of 0.01 nJ. By doing so, all LIF channels are energy calibrated. Those channels included wavelengths at 247, 258, and 396 nm. Substitution of the 416-nm channel (Hanscom) by the 396-nm channel (Otis) was due to a different gas fill in the Raman Shifter. It was difficult to adjust the excitation energies of the weak channels to significantly different levels. As a result, only two or three data points could be collected for each weak channel compared to five data points collected for strong channels on probes A, C, and D and 10 points for strong channels on probe B.

3.3.4 Soil Sample Collection and Analysis

Soil sampling was concentrated in the contaminated area identified by LIF-EEM data at CY4 (Figure 17). The methods used for CPT soil sampling were the same as those used at Hanscom AFB (Section 3.2.3). Two sampling pushes were conducted, and soil samples were collected from various depths at each location corresponding to a variety of LIF detection signals, ranging from nondetect to strong fluorescence. Soil samples were collected from six depths (up to 24 ft) at one location (CY4S), whereas only four samples were collected from the other (CY4T; up to 15 ft) due to the resistance from cobbles.

The procedures used for LIF measurements and the analytical methods for laboratory analyses (Groundwater Analytical, Buzzards Bay, MA) were also the same as those used at Hanscom (Section 3.2.2). The on-site LIF-EEMs collected from the soil samples were analyzed using EEM methods for 10 targeted PAHs, which were found by the analytical laboratory. The LIF-EEM data were analyzed using a combination of abstract factor analysis, target factor analysis, and rank annihilation factor analysis methods previously described in Section 2.6.

To validate the LIF-EEM results, the EEM data were compared to the laboratory data. The in situ LIF data were analyzed for these same compounds and compared to laboratory data from soil samples collected 2 to 3 ft away. A simple one-dimensional measure of the fluorescence fingerprints (total fluorescence summed over all 10 channels) was also correlated with laboratory results for TPH as a way to monitor the results of the LIF measurements in real time. The detailed results of the Otis field test are shown in Section 4.3.2.

3.3.5 Summary of Equipment and Software Performance

Field test No. 2 incorporated equipment and methods improved after the Hanscom field test, provided a demonstration of the LIF-EEM technique, and identified areas for further equipment improvement. A significant amount of data (about 2500 files) were obtained from field operations. Qualitative and semiquantitative analyses were achieved on subsurface soil contaminants, and characterization was completed on a contaminated area at site CY-4. A summary of instrumentation and software performance is presented subsequently.

The improvements incorporated into the LIF-EEM system after the analysis of its performance at Hanscom AFB include the following:

- Speed of data acquisition was increased from 10 to 2 seconds per depth measurement.
- New spectrograph grating was optimized to detect first-order fluorescence signals and eliminate the mixed-order signals detected in Hanscom field data.
- Baseline structure in the fluorescence data was reduced by grounding the outer cover of the CCD with a Faraday cage .
- Pretreatment of EEM data by removal of baseline and backscatter light and energy normalization was streamlined.
- Factor analysis methods were developed and used to analyze in situ and on-site LIF-EEM data.

Instrument operation was generally routine during field test No. 2, although uneven laser performance and absence of a suitable AC power source slowed the field operations. Routine alignment and optimization procedures were performed between and occasionally during soundings. It was found that no conventional gasoline-powered portable generators could provide sufficiently clean or robust power to allow stable laser operation. Instrumentation was again operated off the CPT vehicle internal generators, and an AC inverter was obtained that could run off a portable generator and provide clean single-phase 110-V AC output to the LIF instrumentation

An attempt was made to perform nontriggered automatic data acquisition. The system was programmed to collect LIF data continuously as the CPT. This procedure enabled EEM acquisitions automatically. The estimated depths recorded with the LIF routine, however, did not correlate well with the actual CPT depths because the estimated depths assumed a fixed rate of probe advancement that was not achieved.

In contrast to the cold temperatures encountered at Hanscom, daytime temperatures during the Otis field work were very warm and often approached 30°C in the CPT truck. Temperatures equaled and sometimes exceeded the specified ambient operating temperature range for the laser, which required operation of the laser power supply with its cover removed to allow air circulation from a portable fan. CCD operation was normal, but the detector could not achieve as low an operating temperature as in the laboratory. Although a temperature compatible with that in the laboratory improves the LIF system performance, it can been concluded from the system performance at Hanscom AFB and Otis ANGB that the LIF system functioned well enough to perform site characterization across a wide range of weather conditions.

Recent system modifications are collectively referred to as "final packaging." Those include thermal and vibrational stabilization of the optical bench, active heat exchange from the laser power supply to the outdoors, automatic sampling triggered by CPT depth gauge readings, and a fully automated data analysis routine to enable real-time processing.

3.3.6 Presentation of Fluorescence Data

EEMs were obtained in the same manner at Otis as at Hanscom AFB. Raw EEM data were collected using SpectraMax for Windows software and exported to Excel where they were subjected to baseline subtraction, backscattered excitation light removal, and photon normalization before being analyzed with factor analysis methods. The LIF-EEMs were analyzed using standard EEMs made in the laboratory with the same instrumentation and methodology to identify and semiquantify the PAH contaminants. The on-site LIF-EEM results were compared to the laboratory analytical results from core samples recovered by CPT. The in situ LIF results were used to characterize site CY-4 at Otis. Total fluorescence signals, obtained in the same manner as at Hanscom, were correlated to the TPH concentrations. The results of EEM analysis are presented in Section 4.3.

CHAPTER 4

Data Interpretation and Evaluation

4.1 GENERAL SCOPE OF DATA ANALYSIS

Field laser-induced fluorescence excitation–emission matrix (LIF-EEM) data were transferred to Microsoft Excel for data manipulation. Data were then subjected to pretreatment including removal of backscattered laser excitation, baseline subtraction, and energy normalization using a streamlined routine that accomplished these procedures at a rate of less than 1 second per file.

To identify and quantify the presence of the standards present in the unknown soil samples, the fluorescence signals of rhodamine B (RhB) measurements (Section 3.3.3) for all standards and unknown samples were scaled to the same level, and the scaling factors were used to normalize the sample fluorescence for individual excitation channels. The effects of variation of excitation energies at the probe, optical fiber attenuation, overlap of the excitation and detection surfaces at the probe/sample interface, and spectrograph imaging were thus taken into account.

To validate LIF-EEM results, soil core samples were collected at both Hanscom U.S. Air Force Base (AFB) and Otis U.S. Air National Guard Base (ANGB), and in situ and on-site LIF results were compared to off-site laboratory gas chromatography/mass spectrometry (GC/MS) methods. At Hanscom AFB, the core samples were also analyzed by on-site fast GC/MS for comparison. Laboratory analyses were completed using standard U.S. Environmental Protection Agency (EPA) methods. LIF data were analyzed with several methods, including (1) correlation of total fluorescence signals to total petroleum hydrocarbon (TPH), (2) identification of a specific class of contaminants with the excitation–emission profile (Hanscom) and target factor analysis (Otis), and (3) quantification of contaminants using fluorescence at a specific emission wavelength (Hanscom) and rank annihilation factor analysis (Otis). Finally, combining all information obtained with LIF-EEM measurements, site characterization maps were constructed for the LIF sites at Hanscom and Otis.

4.2 HANSCOM AFB DATA ANALYSIS RESULTS

Total fluorescence is obtained by summing the fluorescence signals at all emission wavelengths from all channels. Figure 18 shows a plot of all detected fluorescence at the Hanscom site. It superimposes the vertical profiles of the contaminants from all CPT locations. It is apparent that the contamination at the site is largely confined to a "hot" zone between the depths of 9 and 12 ft bgs, immediately above the normal level of the water table (i.e., at about 12 to 14 ft, according to observations at nearby recovery wells).

Total fluorescence also provides information about the contaminant at each push location. Figures 19 and 20 are two examples of the in situ contamination profile at locations MWZ10a and

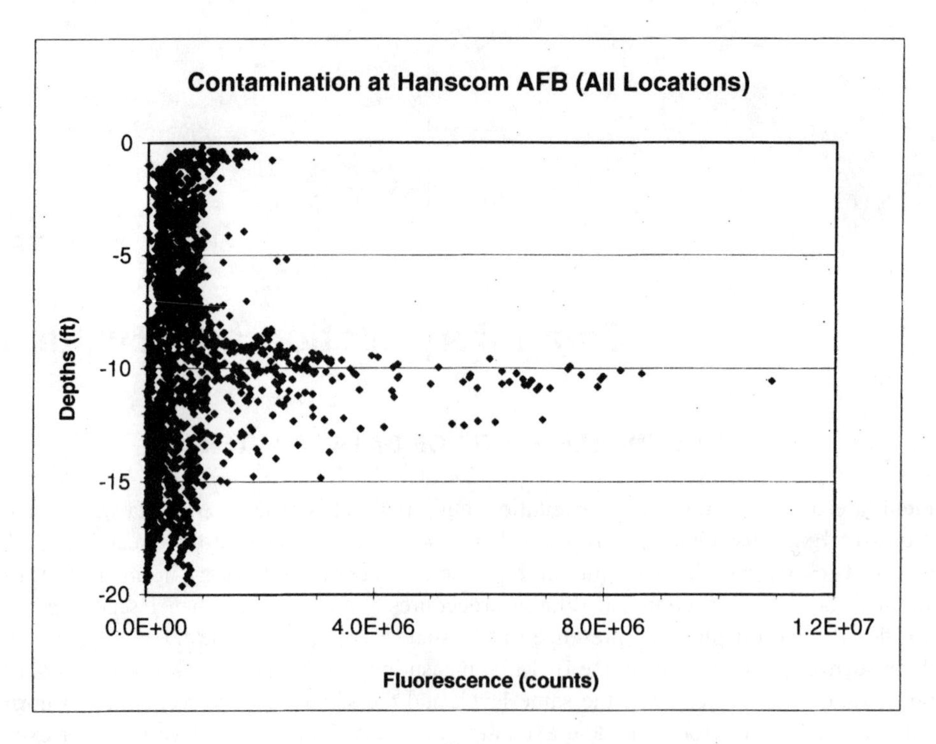

Figure 18 Vertical contamination profiles from all LIF push locations at Hanscom AFB.

RW7b (see Figure 13 site map), respectively. The figures demonstrate the variation of fluorescence, which correlates to TPH concentrations, at a vertical resolution of 2 cm. It is observed that at a depth of about 9.5 ft TPH concentrations start to rise, with data points gradually increasing, and peaking at a depth of about 10 ft followed by a gradual decrease at greater depths. The estimated standard deviations of the LIF measurements are also shown in these two figures.

Some in situ LIF-EEMs are shown in Figure 21. Although all the EEMs were taken at locations around Building 1843 (see Figure 13), it is seen that EEMs taken at locations R7a, R7b, R7c, M9a, R6a, and M10a show a dominant naphthalene fluorescence peak (near 340 nm), whereas those taken at locations M11a, M11b, and M11c have fluorescence from both naphthalene and a stronger presence of heavier polynuclear aromatic hydrocarbons (PAHs) (peak emission at 400 to 425 nm) as well. Comparing these data to standard EEMs of JP-4 and No. 2 fuel oil obtained from the Hanscom site suggests that the contaminated soil represented in the first group of LIF locations contains mainly JP-4, whereas soil at the second group of LIF locations contains both JP-4 and fuel oil.

To validate these LIF results, core samples were analyzed on-site by ex situ LIF and GC/MS and off-site by laboratory analysis. The results from these three methods were compared. Linear correlations were obtained between the LIF signals (i.e., counts) and the off-site laboratory data (i.e., concentrations). The correlations were used as a calibration method to convert the on-site and in situ LIF-EEM results to concentration scales.

For LIF data, the fluorescence signals at 340 nm were analyzed and compared to the laboratory data of naphthalene and diesel range organics (DROs). A linear correlation was obtained for the fluorescence counts, mainly from exciting at a wavelength of 299.1 nm, and the concentrations determined by the laboratory. The linear relationship was used to convert the fluorescence signals to concentrations (i.e., ppm). The photon-normalized total fluorescence signals were correlated to the concentrations of TPH in the sample and converted to ppm in the same manner. The results of naphthalene, TPH, and DRO comparisons among the three methods are shown in Table 4.

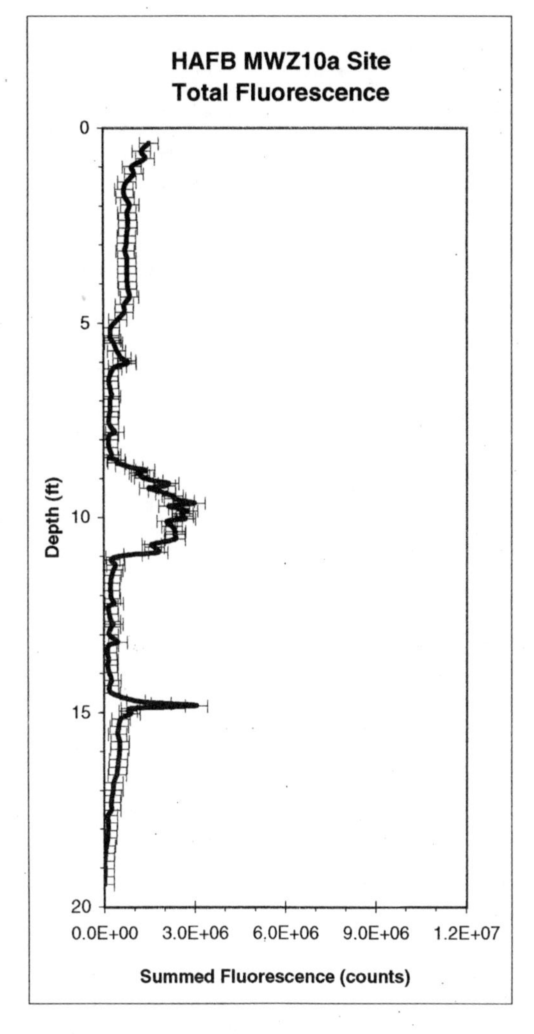

Figure 19 Total fluorescence at Hanscom AFB site MWZ10a.

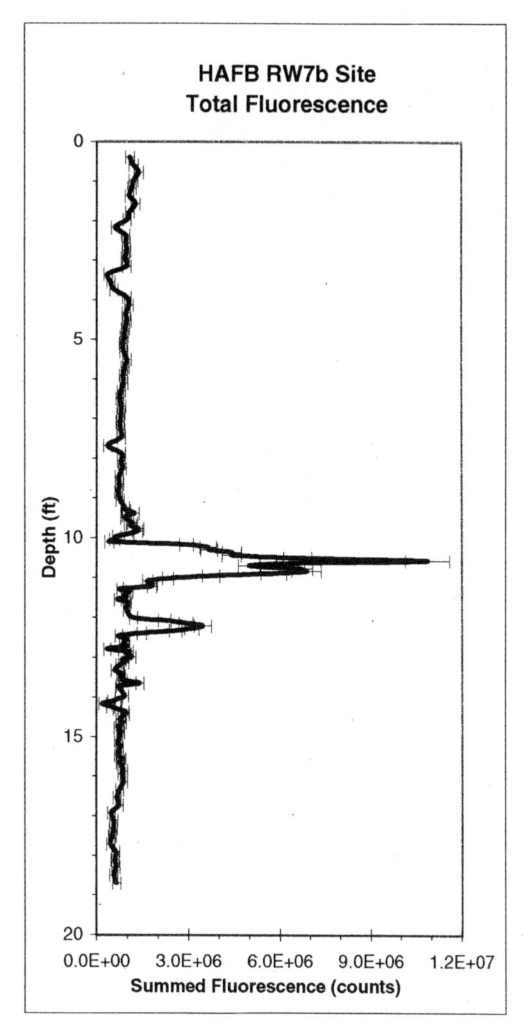

Figure 20 Total fluorescence at Hanscom AFB site RW7b.

Of the 24 samples analyzed for TPH, LIF-EEM gave two false-negative and one false-positive response. In addition, in six cases neither the laboratory nor the LIF detected TPH. For the 15 cases where both LIF and laboratory showed detected signals, LIF showed semiquantitative detection in 11 cases. LIF detected TPH semiquantitatively in 71% of the cases. Similar LIF detection was achieved for DRO, with 70% of the detection within a factor of two. The TPH results of the core samples at the five locations are shown in Figure 13 and are also compared to the in situ LIF signals obtained from the nearby three locations in Figure 22. Good agreement was observed between on-site laboratory results and LIF measurements are observed. Both on-site and in situ LIF data indicated that the most contaminated zone occurs between the depths of 9 to 11 ft bgs.

From analysis of these results, it was determined that the fluorescence signal at the single emission wavelength did not provide enough information to fully classify and quantify chemicals. More rigorous methods utilizing the full EEMs were later developed for identification and quantification purposes. Nonetheless, the information obtained using the TPH correlation was sufficient for mapping the contaminant profile at the Hanscom site. A contaminant distribution map of the Hanscom site was prepared and included as Figure 23, which provides a three-dimensional cross-sectional view of the site. A plane view of the site surface is presented in Figure 24. These figures

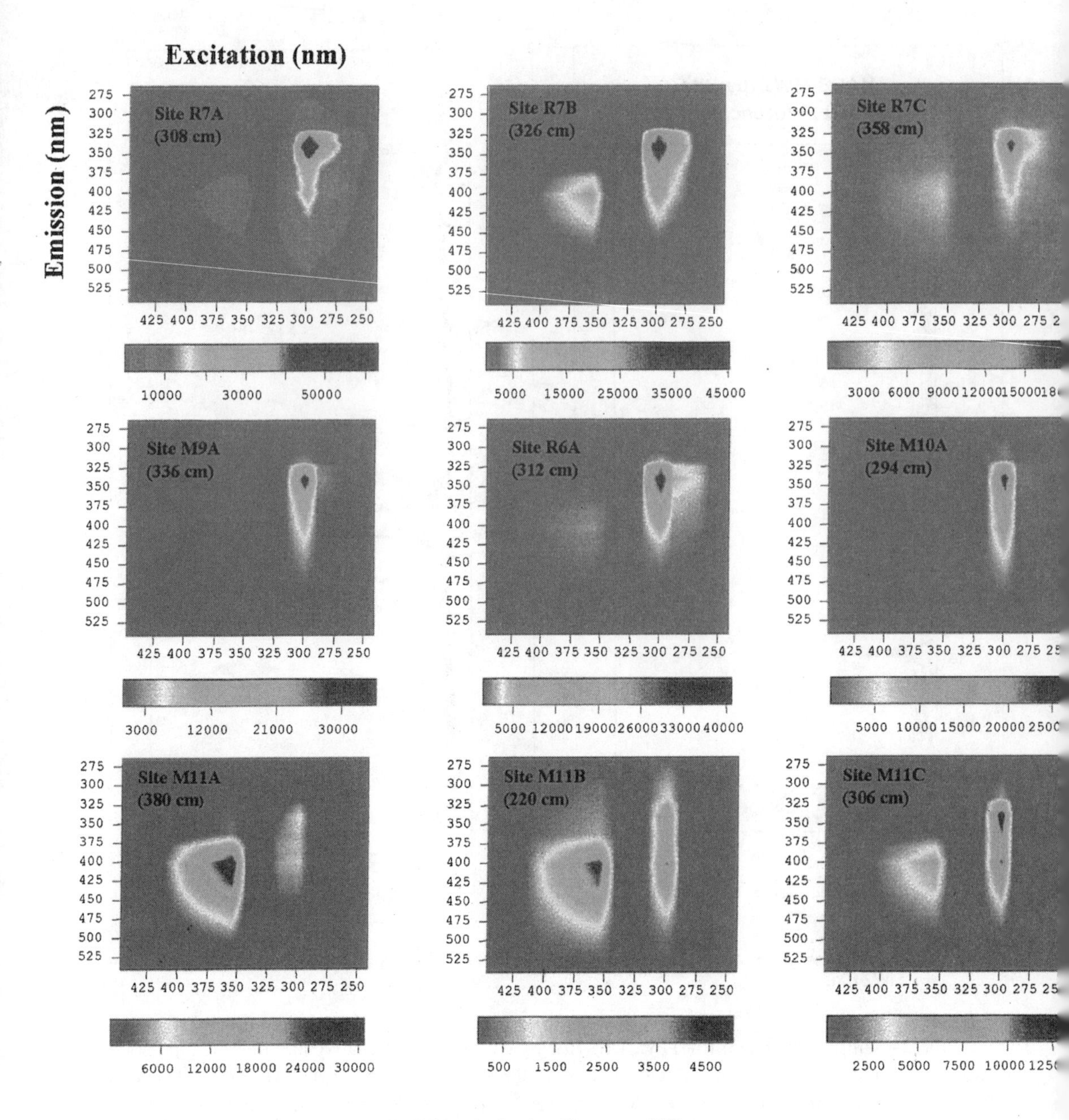

Figure 21 Examples of in situ LIF-EEMs obtained at Hanscom AFB.

were generated with SiteView™ software. The boundary and the profile of the contaminants were constructed by interpolating the in situ LIF signals obtained at the 26 push locations.

4.3 DATA ANALYSIS FOR OTIS FIELD OPERATION

A more rigorous EEM data analysis method was developed after the Hanscom field test. For Otis ANGB data, EEMs were analyzed with a unique factor analysis procedure that consists of three parts: abstract factor analysis (AFA), target factor analysis (TFA), and rank annihilation factor analysis (RAFA).[28] AFA, also called principal component analysis in the literature, was used to

Table 4 Comparison of On-Site and Off-Site GC/MS and LIF-EEM Results

Sample ID (depth, cm)	Target Compounds	Off-site GC/MS (ppm)	On-site GC/MS, (ppm,[a] ave. ± RSD[b])	On-site LIF-EEM (ppm)
R7A (215–245) A	Naphthalene		ND[c]	1
B	Naphthalene		3	ND
C	Naphthalene	3	2, 2, 3 (2.3 ± 25%)	4
	TPH	7700		5400
	DRO	5000		3000
R7A (260–290) A	Naphthalene		2	2
B	Naphthalene		2	ND
C	Naphthalene	3	2, 2, 3 (2.3 ± 25%)	2
	TPH	9300		2800
	DRO	5700		5800
R7A (350–380) C	TPH	41		ND
R7B (225–255) C	TPH	240		ND
	DRO	230		ND
R7B (305–335) A	Naphthalene		ND	ND
B	Naphthalene		ND	2
C	Naphthalene	2	ND, ND, ND	2
	TPH	3200		3500
	DRO	2600		1200
R7E (260–290) A	Naphthalene		4	5
B	Naphthalene		5	ND
C	Naphthalene	5	8, 5, 7 (6.7 ± 23%)	8
	TPH	8900		19,500
	DRO	5800		6000
R7E (305–335) C	TPH	800		500
	DRO	560		ND
R7Y (215–245) A	Naphthalene		3	9
B	Naphthalene		4	3
C	Naphthalene	4	6, 4, 5 (5 ± 20%)	2
	TPH	5300		4500
	DRO	2900		1500
R7Y (260–290) A	Naphthalene		4	6
B	Naphthalene		5	7
C	Naphthalene	5	3, 7, 6 (5.3 ± 39%)	4
	TPH	3500		5600
	DRO	2600		2800
R7Y (305–335) A	Naphthalene		ND	ND
B	Naphthalene		ND	ND
C	Naphthalene	2.4	ND, ND, ND	ND
	TPH	2300		4900
	DRO	2200		250
R7Y (350–380) C	TPH	87		900
	DRO	120		ND
R7Z (100–130) C	TPH	45		ND
R7Z (215–245) A	Naphthalene		8	1
B	Naphthalene		3	3
C	Naphthalene	5	7, 7, 6 (6.7 ± 9%)	3
	TPH	7600		3800
	DRO	4000		2300

Table 4 Comparison of On-Site and Off-Site GC/MS and LIF-EEM Results (Continued)

Sample ID (depth, cm)	Target Compounds	Off-site GC/MS (ppm)	On-site GC/MS, (ppm,[a] ave. ± RSD[b])	On-site LIF-EEM (ppm)
R7Z (260–290) A	Naphthalene		ND	ND
B	Naphthalene		3	ND
C	Naphthalene	5	2, 3, 2 (2.3 ± 25%)	3
	TPH	5400		5100
	DRO	2900		2000
R7Z (305–335) A	Naphthalene		4	19
B	Naphthalene		3	3
C	Naphthalene	12	12, 6, 4 (7.3 ± 57%)	16
	TPH	18,000		18,800
	DRO	7500		11,500

[a] The average concentration was reported for all C naphthalene samples by field GC/MS (N = 3).
[b] RSD, relative standard deviation.
[c] ND, no measurable amounts detected.

determine the number (n) of the fluorescent components in the unknown mixture, TFA was used to identify these components, and RAFA was used to quantify each component by determining the unknown/standard concentration ratio. The n values obtained from AFA were subsequently used in the TFA and RAFA methods. All of the computer programs used in factor analysis were written in MATLAB™ language and were run under MATLAB 5.0™. The combination of the factor analysis routines yielded information on both the identity and quantity of the fluorescent components in the unknown mixture.

4.3.1 LIF Calibrations and Tests Performed in Tufts Laboratory

4.3.1.1 Individual-Component Standard Solutions

The laboratory data from Groundwater Analytical indicated the presence of 10 semivolatile organics, including anthracene (A), benzo(a)anthracene (Ba), benzo(b)fluoranthene (Bf), benzo(a)pyrene (Bp), chrysene (C), fluoranthene (F), 2-methyl-naphthalene (M), naphthalene (N), phenanthrene (P), and pyrene (Py). *p*-Xylene (X) was also found in the preliminary site characterization conducted in March 1996. These 11 chemicals were therefore chosen as standards for EEM factor analysis. Solutions of each of the chemicals were made in cyclohexane. For each chemical, a sand standard was made by saturating sand with the solution. The sand used was taken from a "clean" location at CY-4 site so that the particle size of the sand was representative of the site soil texture (sand and silty sand, determined by the particle size analysis conducted by Groundwater Analytical) for the in situ LIF-CPT measurements. No background fluorescence was detected from the sand alone. The concentrations of the standard solutions are shown in Table 5.

The solutions above yielded raw fluorescence at levels of thousands or tens of thousands of counts. It was found that when the concentration of a sample was very low, such as in the case of sand standard samples, the signals from some weaker channels were reduced to the level of background noise. As a result, the solution EEMs were used as standards for the data analysis. Because a RAFA result consists of a ratio of concentration of a target analyte in the sample to its concentration in the standard, the use of neat solutions as standards required multiplication of the RAFA ratio by an appropriate conversion factor (discussed subsequently) to obtain the concentration of the sample in mg/kg of soil.

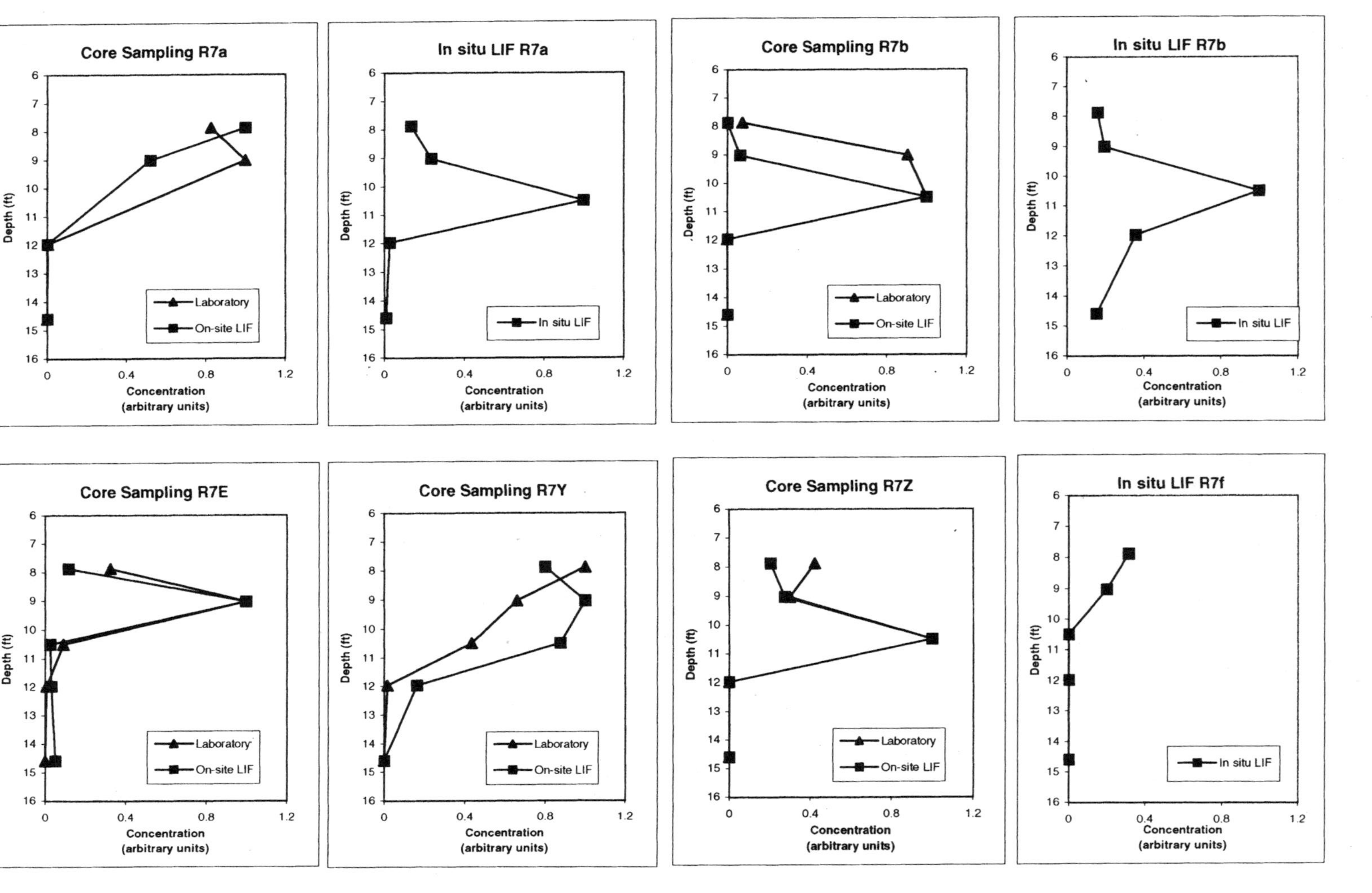

Figure 22 Comparison of results from core sampling and from in situ LIF measurements.

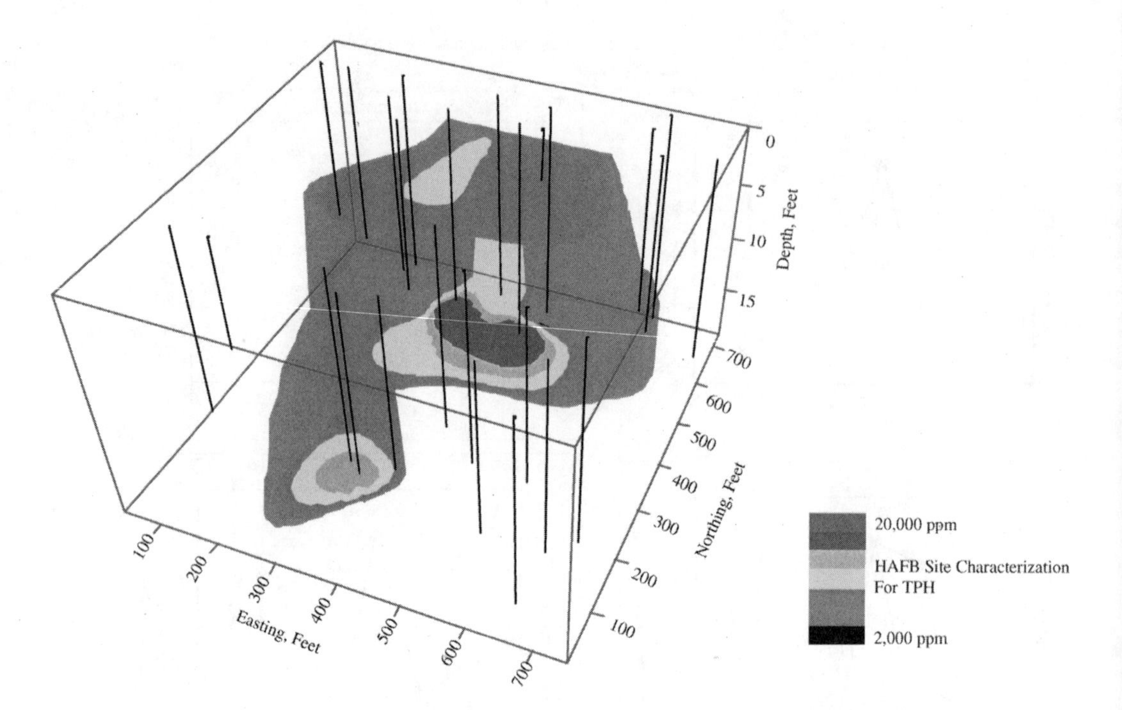

Figure 23 Hanscom Air Force Base site characterization (three-dimensional visualization).

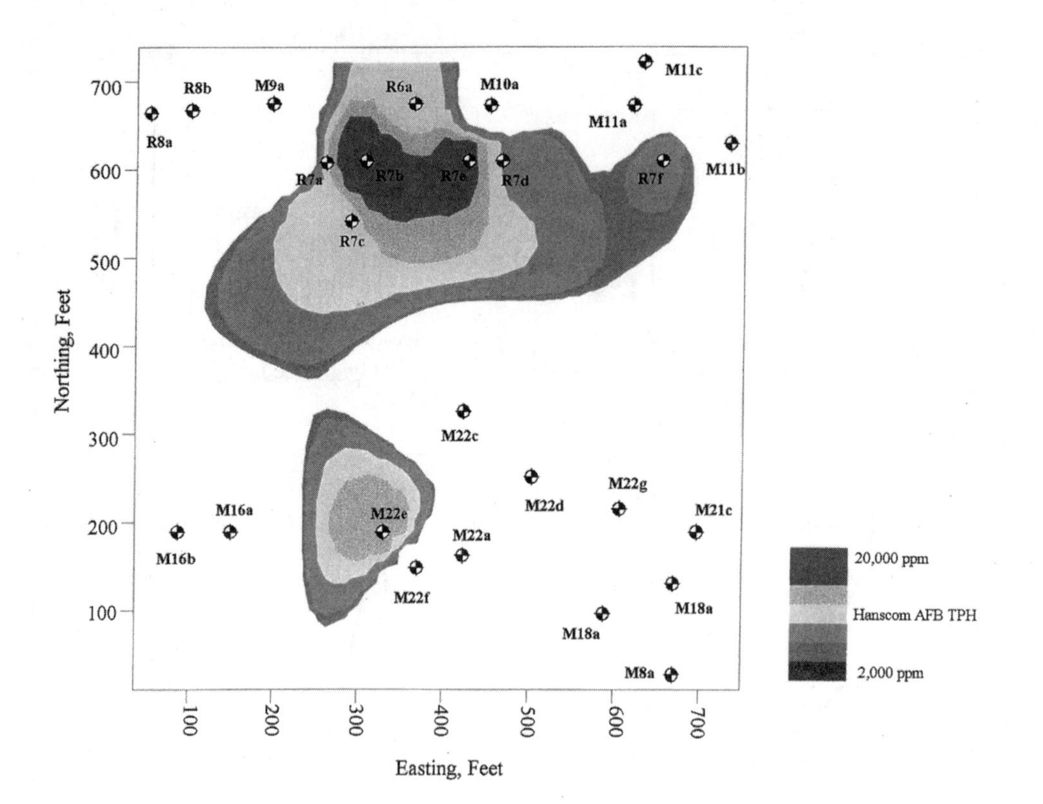

Figure 24 Hanscom Air Force Base site characterization (two-dimensional visualization).

Table 5 The Concentrations of Standard Solutions for Factor Analysis

Name	Component	Solution Concentration (mg/l)
A	Anthracene	7.50
Ba	Benzo(a)anthracene	6.50
Bf	Benzo(b)fluoranthene	12.00
Bp	Benzo(a)pyrene	2.86
C	Chrysene	10.46
F	Fluoranthene	11.50
M	2-Methyl-naphthalene	13.60
N	Naphthalene	14.00
P	Phenanthrene	10.00
Py	Pyrene	7.50
X	*p*-Xylene	200.00

Detection of PAHs in Classes — The value of this data analysis strategy is that a sample could be analyzed for any target compound or class of compounds in the database, and its concentration in the sample could be determined independently from the presence or absence of any other fluorescent species. The ability to discriminate among these 11 standard compounds was determined first. Using each standard as an "unknown" and analyzing it for the presence of each of the 11 compounds in the database, factor analysis correctly identified each sample, with the exception of the chemically and spectroscopically similar naphthalene and 2-methylnaphthalene, that is, the 11 standards were detected as 10 classes, with naphthalene and 2-methylnaphthalene detected together as one class. The ability to discriminate among all other species in the previous list was encouraging, given the strong similarities between some chemical structures. It was expected that very similar chemicals would be detected generically as a class, given the limited resolution of the excitation spectra provided by only the wavelengths that were selected.

In the case of multiple analytes being detected as a class instead of individually, the LIF-EEM analysis described previously did not yield absolute quantitative information about the concentrations of each or the total concentration of all members of the class. The technique generally has a different sensitivity to each member of the class, depending upon its absorption strength and fluorescence quantum yield. However, it can be used to obtain a total measure of all members in the class, expressed as equivalents of some particular member of the class. Such reporting protocol is conventional with many analytical techniques that are not completely molecule-specific. It is found that the LIF instrument response (i.e., sensitivity factor) for naphthalene is approximately half of that for 2-methylnaphthalene. This sensitivity factor must be utilized in comparing laboratory results for individual species with the LIF-EEM analysis results.

Conversion from LIF-EEM/Factor Analysis to Concentration in Soil — When the physical state of the sample used to measure the standard EEM of component A, $EEM_{std,A}$, is the same as the physical state of the unknown sample whose EEM_{sample} is being analyzed, the RAFA multiplier, x_A in Equation 1 simply represents the ratio of concentrations of A in the sample and in the standard. Therefore, the concentration of A in the sample, $C_{A,sample}$ is equal to $x_A C_{A,std}$. However, in this case, an additional conversion factor, f, must be included to account for differences in the sample (measured in situ in soil) and the standard (measured in cyclohexane solution).[28] This conversion factor is included in Equation 2:

$$C_{A,sample} = f x_A C_{A,std} \tag{2}$$

If sample conditions can be duplicated in the laboratory by preparing mixtures of fluorescent analytes with clean soil from the site f may be determined experimentally. This was done for the target analytes listed previously, and the results indicated that the value of f was species dependent. This dependency arises from the fact that the fluorescence from each species is distributed differently among the excitation channels, and the channels themselves have different responses because of slight geometry differences from machining or assembly at the probe. For naphthalene and 2-methylnaphthalene, however, the f factors should be the same because of the similarities in their excitation and emission profiles. This was verified with additional measurements on a conventional fluorimeter.

Sensitivity Factors — The sensitivity factors for these two components, naphthalene and 2-methylnaphthalene, were obtained using one as the standard and analyzing the other as the "unknown." The results indicated that LIF-EEM measurement for 2-methylnaphthalene was approximately twice as sensitive as that for naphthalene. Thus, when a sample containing both compounds is analyzed for the naphthalene class ($\boldsymbol{n}$), expressed as equivalents of naphthalene ($\boldsymbol{n}_{\mathrm{LIF,\,N}}$), the result obtained may be expressed as

$$\boldsymbol{n}_{\mathrm{LIF,\,N}} = k\,(N + 2.390M) \tag{3}$$

whereas the result expressed as equivalents of 2-methylnaphthalene would be given by

$$\boldsymbol{n}_{\mathrm{LIF,\,M}} = k\,(0.490N + M) \tag{4}$$

where N and M are the actual concentrations of naphthalene and 2-methylnaphthalene, respectively, and k is a constant depending on the sample characteristics (e.g., particle size, relative saturation, etc.). A k value of 0.110 was obtained from multiple-regression fits of LIF vs. laboratory data for the on-site soil measurements.

4.3.1.2 Multicomponent Solution Standards

To test the validity of the factor analysis methods, nine solution mixtures were prepared with two to five components each. EEMs of these mixtures were collected using the LIF-CPT probe for the on-site LIF measurements at Otis ANGB and subjected to factor analysis. The compositions of these mixtures are listed in Table 6.

In Table 6, the column "Concentration (mg/l)" contains the absolute concentrations of the solution mixtures. Data in the "Concentration Ratio/Real" column are calculated as the ratio of the absolute concentration of the mixture over the absolute concentration of the single-component standard (Table 5). For the NM sample, the real ratio for naphthalene is 5.73/14.00 = 0.41 and that for 2-methylnaphthalene is 5.56/13.6 = 0.41. Data in the "Concentration Ratio/Measured" are calculated as the ratio obtained by rank annihilation factor analysis (RAFA), which is the ratio of the EEM intensity of the component in the mixture over the EEM intensity of the single-component standard. For NM, the ratio of EEM intensity of N in the mixture over the EEM intensity of N standard is 0.92, and the ratio of EEM intensity of M in the mixture over the EEM intensity of M standard is 0.58. Because the EEMs are analyzed against two different standards, there are two ratios obtained, each corresponding to a specific standard.

RAFA provides a ratio for a component that it "thinks" is present in the mixture, which is not always true. In the case of NM and NMP, both naphthalene and 2-methylnaphthalene are present, so it does not seem problematic that it provides the ratios for both compounds. However, with single-component naphthalene as the "unknown mixture" and 2-methylnaphthalene as the standard, RAFA still thinks that 2-methylnaphthalene is present and gives a ratio that indicates the 2-methylnaphthalene concentration. This also happens with the reverse (i.e., 2-methylnaphthalene as unknown mixture and naphthalene as standard). It was concluded that the two components are not separable with the current EEM method, and that the two components are detected as a group.

Table 6 Composition of Laboratory Solution Mixtures and Mixture/Standard Concentration Ratios from Factor Analysis

Sample Name	Component	Concentration (mg/l)	Concentration Ratio	
			Real	Measured
NA	Naphthalene	7.96	0.57	0.55
	Anthracene	1.70	0.23	0.23
NAF	Naphthalene	7.96	0.57	0.45
	Anthracene	1.70	0.23	0.24
	Fluoranthene	1.31	0.11	0.42
NM	Naphthalene	5.73	0.41	0.92
	2-Methyl-naphthalene	5.56	0.41	0.58
NMP	Naphthalene	3.98	0.28	0.95
	2-Methyl-naphthalene	3.70	0.27	0.53
	Phenanthrene	1.82	0.18	0.22
NAFX	Naphthalene	7.96	0.57	0.54
	Anthracene	1.70	0.23	0.22
	Fluoranthene	1.31	0.11	0.22
	p-Xylene	455	2.28	1.67
Ppy	Phenanthrene	1.82	0.18	0.19
	Pyrene	2.73	0.36	0.51
PpyBf	Phenanthrene	1.67	0.17	0.17
	Pyrene	2.5	0.33	0.48
	Benzo(b)fluoranthene	1.0	0.08	0.15
PpyBfBa	Phenanthrene	1.54	0.15	0.14
	Pyrene	2.31	0.31	0.38
	Benzo(b)fluoranthene	0.92	0.08	0.11
	Benzo(a)anthracene	0.50	0.08	0.22
PpyBfBaX	Phenanthrene	1.54	0.15	0.62
	Pyrene	2.31	0.31	0.46
	Benzo(b)fluoranthene	0.92	0.08	0.12
	Benzo(a)anthracene	0.50	0.08	0.17
	p-Xylene	384	1.92	1.67

In Table 6, although there are two ratios obtained for N and M, there is a problem with the accuracy of the RAFA results for these two components. The absolute concentration ratio of naphthalene and 2-methylnaphthalene should be about 1 to 1 (0.41 vs. 0.41 in NM and 0.28 vs. 0.27 in NMP), but the measured ratio of these compounds is almost 2 (for naphthalene) to 1 (for 2-methylnaphthalene), which are both higher than the real values. This occurs because RAFA cannot separate these two components, and each of them contributes to the EEM signal of the other. Equations 3 and 4 are implemented to account for this sensitivity difference.

Results of AFA — All nine mixtures were analyzed first by AFA to determine the number of factors (n), or number of the principal components. For the solution standards, AFA methods correctly determined the number of the components for all nine mixtures (e.g., in Table 6, AFA result for a mixture of NAFX showed $n = 4$, which indicates that four components are present in the mixture).

Results of TFA — The n value data obtained from AFA are sequentially used in TFA and RAFA methods. For the 11 standards, individual components in the 9 mixtures were correctly identified by TFA. This is an important step for data analysis because in some cases (especially with data obtained in the field), the traditional rank annihilation analysis methods (i.e., used without TFA) might give positive concentration ratio results for a certain component although the component is not present.

RAFA Results — The concentration ratios of unknowns/standards for the components identified by the TFA analyses were calculated using RAFA, with knowledge of the principal factor (n) values determined by AFA. All of the concentration ratios for the components in the nine mixtures determined by RAFA are listed in Table 6. The agreement with the real ratios is reasonably good, with 23 of 28 results (and 2 additional results near) within a factor of 2 (the limits defined for "semiquantitative" results) between the real value and the value from RAFA. Exceptions occur for naphthalene in the mixtures of NM and NMP, fluoranthene in NAF, benzo(a)anthracene in PPyBfBa, and phenanthrene in PPyBfBaX.

The relatively poor agreement for both naphthalenes in NM and NMP occurs because naphthalene (N) and 2-methylnaphthalene (M) are not distinguished, and both substances contribute to the apparent concentration of the other, according to their sensitivity factors as previously discussed. Consequently, the naphthalene concentration (ratio × concentration of standard solution) detected by LIF should be compared to the weighted sum of laboratory-determined concentrations, $n_{LIF,\,N} = N + 2.039\,M$, and the 2-methylnaphthalene LIF result should be compared to the weighted sum, $n_{LIF,\,M} = M + 0.490\,N$, as indicated in equations 3 and 4 ($k = 1$ in solution), and the results are shown in Table 6a.

Table 6a Continued from Table 6, but Naphthalene and 2-Methylnaphthalene Concentrations are Weighted

Sample Name	Component	Concentration (mg/l)	Concentration Ratio	
			Real	Measured
NM	Naphthalene	17.07	1.22	0.92
	2-Methyl-naphthalene	8.37	0.62	0.58
NMP	Naphthalene	11.52	0.82	0.95
	2-Methyl-naphthalene	5.65	0.42	0.53
	Phenanthrene	1.82	0.18	0.22

When these comparisons are performed, agreement between the LIF data and the weighted-sum improves significantly for both N and M in both mixtures (errors less than 25% in all cases). The fraction of the quantitative results that are correct to within a factor of 2 also improves to 25 of 28, or 89%.

4.3.2 Results of On-Site Soil Sampling

The following sections refer to both "on-site" and "in situ" LIF data. On-site refers to data collected by the LIF instrument on soil cores collected for laboratory analysis. LIF measurement is nondestructive, so laboratory and LIF results are available on the same soil samples. Comparison of these results can provide an assessment of the technique. In situ refers to LIF data collected with the probe deployed in the subsurface during a CPT sounding. These data cannot be compared directly to any laboratory results because soil samples cannot be extracted from the same push location used for LIF and because the levels of contamination were very localized. The concentrations of contaminants changed significantly for adjacent pushes separated by only a few feet. Laboratory results can be compared for the two soil sampling pushes, S and T, with the results of one adjacent LIF push, J, in the next section, to indicate general agreement between the on-site and in situ LIF data sets.

Figure 16 shows the location of the test site at Otis ANGB, and Figure 17 indicates the locations of sampling for on-site and in situ LIF-CPT data collection. Soil cores were collected using the CPT soil sampler at two locations at site CY-4 (CY4S and CY4T, Figure 17). After on site LIF measurements were collected, the samples were sent to Groundwater Analytical Laboratory for analysis of semivolatiles and TPH using standard EPA methods. Data obtained by these two methods were then compared. From the laboratory analyses, it was observed that naphthalene and

2-methylnaphthalene were present in all core samples and were the major components at the site, with relatively high concentrations (up to 41 mg/kg). Other semivolatiles, such as anthracene, fluoranthene, pyrene, and chrysene, were only detected in one or two samples.

As mentioned in the previous section, solutions of individual PAH species in cyclohexane, indicated in Table 5, were used as standards for factor analysis of the on-site LIF data. Species-dependent conversion factors, f_A, were used to obtain approximate contaminant concentrations in mg/kg. For naphthalene and 2-methylnaphthalene, these factors were obtained by optimizing the agreement between off-site laboratory and on-site LIF-EEM data from core samples. The other analytes were determined using soil from the site in laboratory measurements as described previously. The results for all the soil core samples are shown in Table 7.

Table 7 Comparison of On-site EEM Analysis Results and Laboratory Results for Field Data

		Lab Results, Concentration		LIF-EEM Results	
Sample	**Components**	**(mg/kg)**	**Reporting Limit**	**Concentration Ratio**	**Concentration (mg/kg)**
s6a	Naphthalene	34.95	0.37	0.033	11.78
	2-Methyl-naphthalene	17.14	0.37	0.016	5.55
s6b	Naphthalene	23.88	1.80	0.012	4.11
	2-Methyl-naphthalene	11.71	1.80	0.006	1.94
s13a3	Naphthalene	6.37	0.34	0.016	5.57
	2-Methyl-naphthalene	3.12	0.34	0.007	2.50
	Benzo(b)fluoranthene	Not found	0.34	0.004	0.89
s13b3	Naphthalene	6.47	0.35	0.050	17.89
	2-Methyl-naphthalene	3.17	0.35	0.027	9.47
	Phenanthrene	0.36	0.35	Not found	Not found
	Benzo(b)fluoranthene	Not found	0.35	0.003	0.53
	Benzo(a)pyrene	Not found	0.35	0.006	0.28
s15b	Naphthalene	26.47	1.70	0.100	35.71
	2-Methyl-naphthalene	12.98	1.70	0.053	18.38
	Fluoranthene	Not found	1.70	0.033	13.32
	Benzo(a)pyrene	Not found	1.70	0.029	1.47
s19a	Naphthalene	28.07	1.70	0.080	28.56
	2-Methyl-naphthalene	13.77	1.70	0.039	13.53
	Fluoranthene	Not found	1.70	0.046	18.56
	Benzo(a)pyrene	Not found	1.70	0.024	1.21
s22a	Naphthalene	10.44	0.34	0.115	41.06
	2-Methyl-naphthalene	5.12	0.34	0.068	23.59
s22b3	Naphthalene	42.66	1.70	0.132	47.13
	2-Methyl-naphthalene	20.92	1.70	0.066	22.89
	Fluoranthene	Not found	1.70	0.015	6.05
	Benzo(a)pyrene	Not found	1.70	0.010	0.51
t3a3	Naphthalene	13.07	0.35	0.038	13.57
	2-Methyl-naphthalene	6.41	0.35	0.019	6.59
	Benzo(b)fluoranthene	0.58	0.35	0.014	0.71
	Phenanthrene	0.63	0.35	Not found	Not found
	Benzo(a)pyrene	0.38	0.35	Not found	Not found
t6a	Naphthalene	105.60	3.70	0.291	103.90
	2-Methyl-naphthalene	51.79	3.70	0.161	55.84
t6b3	Naphthalene	61.94	1.80	0.186	66.41
	2-Methyl-naphthalene	30.38	1.80	0.082	28.44
t11a3	Naphthalene	4.08	0.34	0.032	11.43
	2-Methyl-naphthalene	2.00	0.34	0.017	5.90
t13a3	Naphthalene	4.49	0.37	0.021	7.50
	2-Methyl-naphthalene	2.20	0.37	0.011	3.82

Note: Laboratory results shown here for naphthalene and 2-methylnaphthalene are weighted according to Equations 3 and 4 to be compared directly to the LIF-EEM results.

(Table 6 listed laboratory mixtures "randomly" chosen to test the accuracy of factor analysis; however, Table 7 lists mixtures from real-world samples, collected on-site at Otis ANGB. There is no direct relationship between these two tables.)

It is seen in Table 7 that the chemical concentrations obtained using these two methods (on-site LIF and off-site laboratory analyses) are within a factor of 2 for 17 of the 25 cases in which quantitative comparison is possible and within a factor of 3 for an additional 6 cases. In only three cases did the LIF fail to detect a species found by the laboratory; in two of these cases the concentration was approximately equal to the detection limit, and in the third it was twice the detection limit. Conversely, among the nine cases where LIF showed detectable signals and the laboratory showed a nondetect, four involved concentrations below the reporting limits of the laboratory method. This may suggest that LIF-EEM can achieve a sensitivity comparable to that of the laboratory method. In two of the cases, LIF found fluoranthene at concentrations that were more than five times higher than the reporting limit of the laboratory methods, whereas the laboratory reported nondetect. This might be related to the inaccuracy of the conversion factor for fluoranthene. In all, 62% of the LIF results fall in the acceptable range for semiquantitative methods (factor of 2 accuracy), and 80% are accurate to within a factor of three.

Other LIF-CPT developers use the results of a conventional measurement on the field sample to determine a calibration factor to convert the LIF readings to an absolute concentration scale.[25] Use of the off-site laboratory data to adjust the f_A values of the two naphthalenes is equivalent to such a calibration. Alternatively, the two sets of results can be plotted vs. depth on a relative concentration scale. This is done for the naphthalene class in Figure 25, where concentrations on core samples as determined by LIF (reported as equivalents of naphthalene) are compared to the properly weighted sums of the laboratory concentrations of naphthalene and 2-methylnaphthalene for the sampling location CY4T. (The CPT probe could only be advanced to a depth of approximately 14 ft bgs at this location because of resistance from cobbles in the soil.) The results from the LIF measurements and the laboratory tests agree well. The LIF and analytical results from cores at the other soil sampling location (CY4S) are shown in Figure 26. At this location, agreement between LIF and laboratory data is less striking, although the general concentration trends correspond. Both data sets indicate low contaminant concentrations at a depth of 13 ft bgs and highest contaminant concentrations near a depth of about 24 ft bgs. Some differences may be expected because of soil inhomogeneity because the LIF sampling occurs over only a small portion of the soil core before it is homogenized in the laboratory prior to extraction and analysis.

4.3.3 Results of In Situ LIF Measurements

4.3.3.1 Comparison of In Situ LIF and Laboratory Results

Because most of the in situ LIF/CPT measurements extended to depths of about 24 ft bgs, the soil sampling at CY4S (which only reached a depth of about 24 ft bgs) is used for comparison to in situ LIF data. The in situ LIF-CPT measurements for comparison were conducted at a location (CY4J) about 2 ft away from the sampling location (i.e., CY4S, see Figure 17). A comparison of the in situ LIF-EEM measurements (naphthalene class, expressed as equivalents of naphthalene) and laboratory results (appropriately weighted sum of naphthalene and 2-methylnaphthalene) from location CY4S is plotted in Figure 27.

General agreement is obtained between these two methods, but the comparison really displays the difference in vertical resolution clearly. The depth resolution for the soil sampling was 2 ft, whereas that for in situ LIF measurement was 2 cm. It is apparent that in situ LIF can provide greater detail (more data points) and, therefore, more accuracy in locating zones of contamination. To compare the quantitative results on an equal-resolution basis, the LIF data were then averaged according to the depths of the soil sampling (i.e., the 30 LIF measurements obtained within the 2-ft range of a given soil core sample were averaged). The comparison results are plotted in Figure 28. The standard

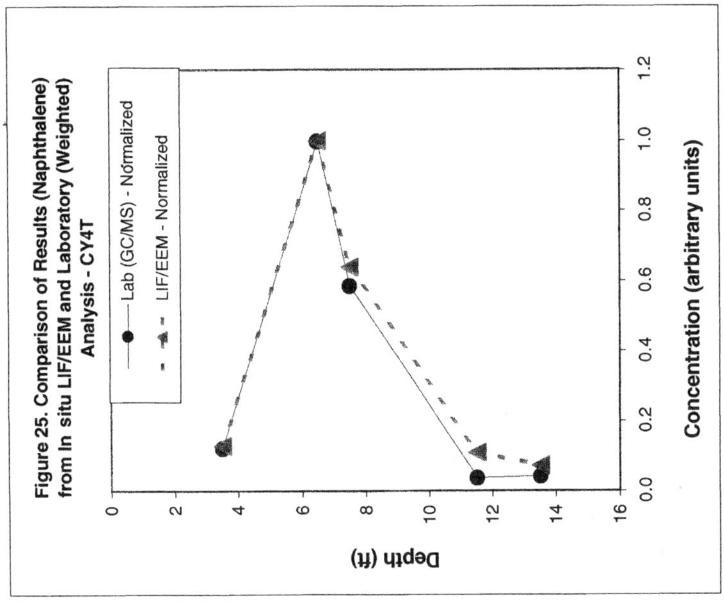

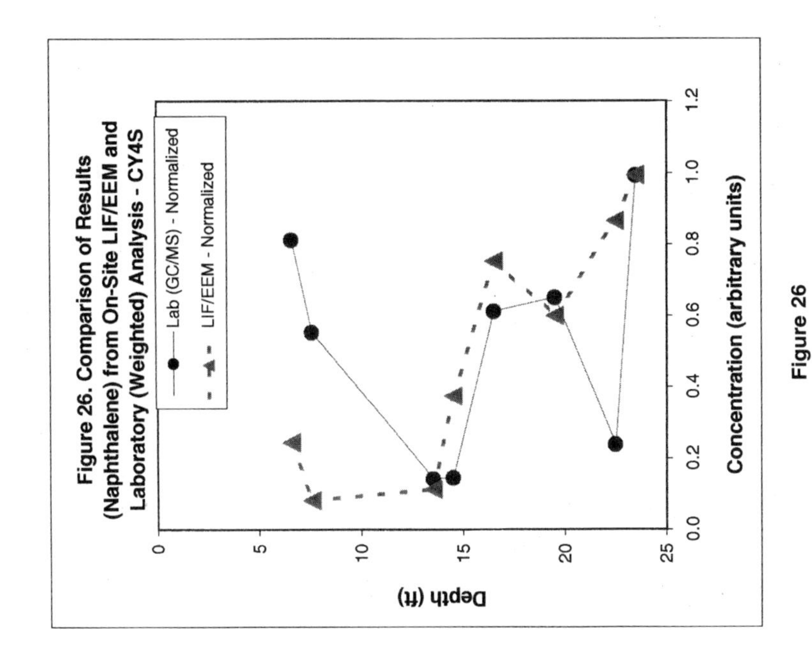

Lab (GC/MS) - Normalized
LIF/EEM - Normalized
Depth (ft)
0
2
4
6
8
10
12
14
16
0.0
0.2
0.4
0.6
0.8
1.0
1.2
Concentration (arbitrary units)

Figure 25. Comparison of Results (Naphthalene) from In situ LIF/EEM and Laboratory (Weighted) Analysis - CY4T

Figure 25

Lab (GC/MS) - Normalized
LIF/EEM - Normalized
Depth (ft)
0
5
10
15
20
25
0.0
0.2
0.4
0.6
0.8
1.0
1.2
Concentration (arbitrary units)

Figure 26. Comparison of Results (Naphthalene) from On-Site LIF/EEM and Laboratory (Weighted) Analysis - CY4S

Figure 26

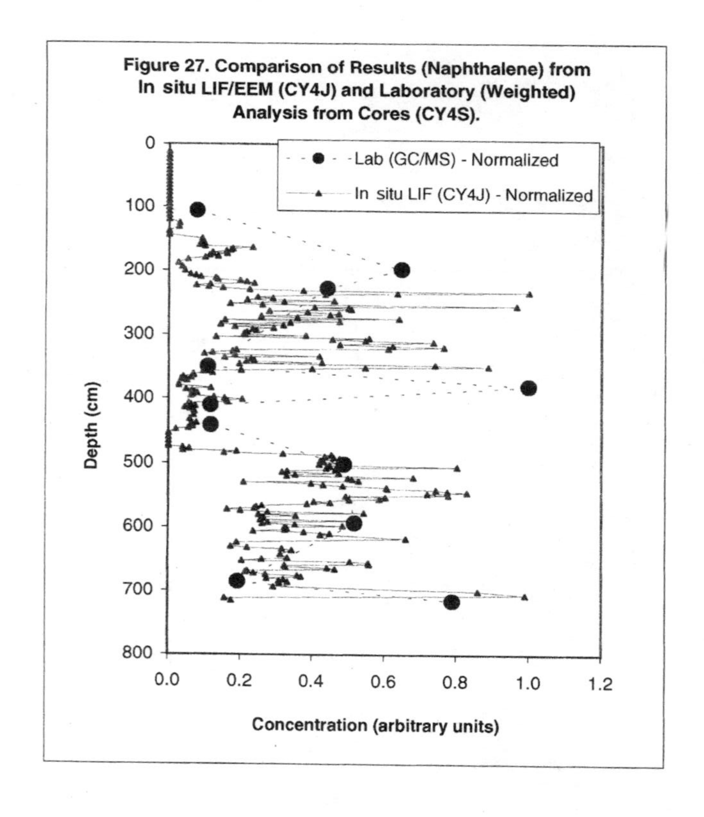

Figure 27

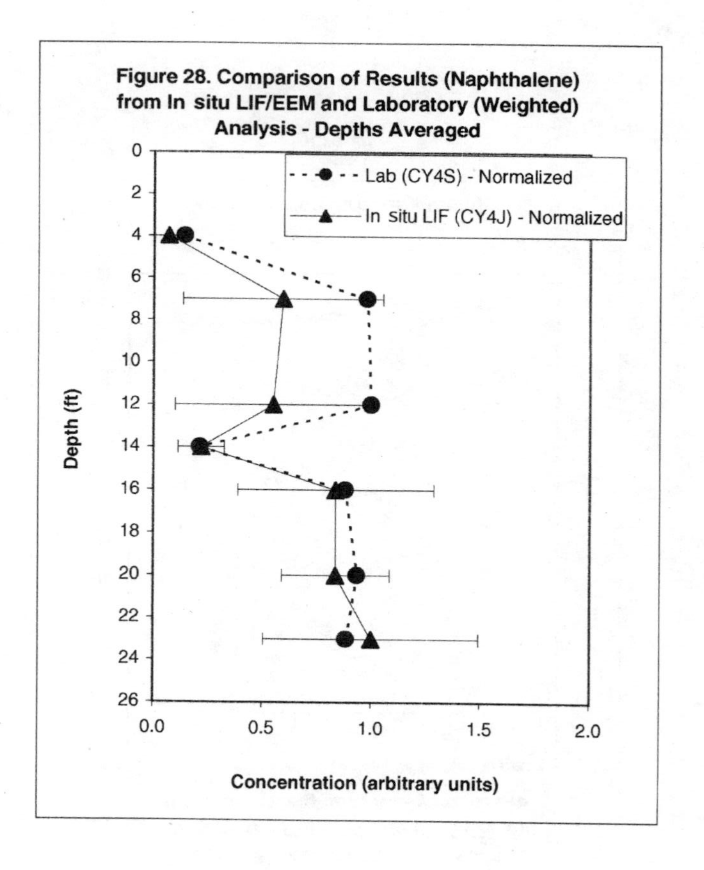

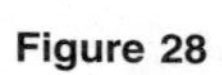

Figure 28

deviation of each of the 30 LIF measurements is also indicated. It can be seen that the laboratory results for location CY4S and the in situ LIF-EEM results for CY4J, 2 ft away, are in good agreement.

4.3.3.2 Site Characterization

LIF-CPT measurements have been used to characterize the CY-4 site at Otis ANGB in Massachusetts. Significant levels of PAH contamination were found at this site localized within an area of about 100 sf^2, with its approximate center at location CY-4J (Figure 17).

As at Hanford AFB, EEM images are plotted for visualization of the contaminants present in the CY-4 site. The EEM profiles at two different depths obtained at the CY-4J location are shown in Figure 29. It is seen that at the shallower depth (i.e., 314 cm) the contaminants are mainly naphthalenes, whereas at greater depth (i.e., 652 cm), the contribution from heavier PAHs, such as fluoranthene, plays a significant role. The detailed mathematical analysis results of the CY-4 site are presented as following.

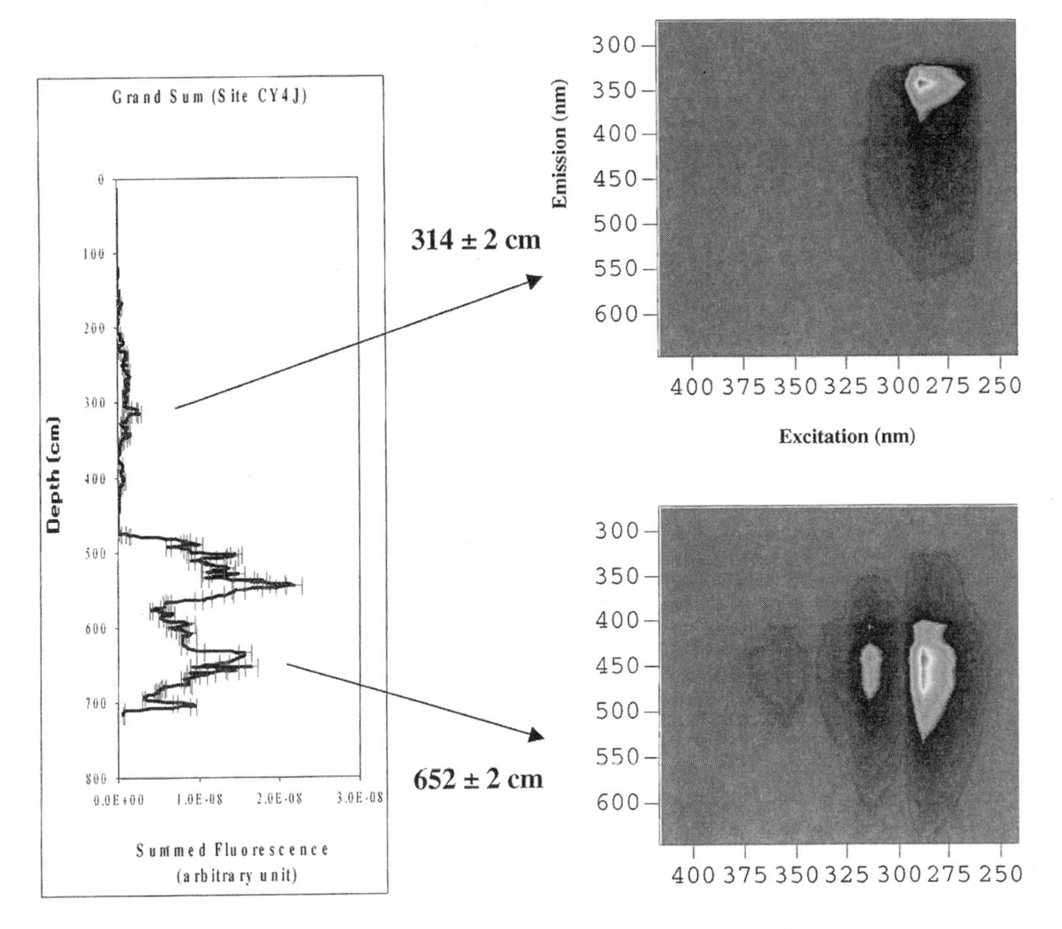

Figure 29 Shallow and deep EEM fluorescence profiles from site CY4J Otis ANGB.

Contaminant Speciation by Classes — As indicated by data from the analytical laboratory, the most prominent contamination at this site, especially around CY-4J, is attributed to naphthalene and 2-methylnaphthalene. Due to the difficulty in differentiating EEMs from these two components, they are grouped, and naphthalene is chosen to represent the contaminants of this class. The LIF-EEM measurements found this class of contamination at six push locations: CY4T, CY4S, CY4J,

CY4P, CY4Q, and CY4M (Figures 30a to 30f). The first two are from on-site soil sampling measurements (with depth resolution of 2 ft), and the last four are in situ LIF data (with depth resolution 2 cm). In Figures 30a to 30f, the relative concentrations of naphthalene (normalized to 1 within each push for easy visual inspection) are shown. In Figure 31, all concentrations are normalized to the contamination scale of CY4J (i.e., the location with the most contaminants) for comparison among the locations.

Factor analysis data indicates that neither naphthalene nor 2-methylnaphthalene was detected at other locations. With the naphthalene and 2-methylnaphthalene profiles for each location, a site characterization map for this class of PAHs, shown in Figure 32, was generated. It is observed that the naphthalene contamination at this site is more or less localized in two horizontal layers, located at 7 to 10 ft and 18 to 22 ft.

In addition to naphthalene and 2-methylnaphthalene, heavier PAHs such as fluoranthene and benzo(b)fluoranthene were also detected at several locations, including CY4C, CY4D, CY4E, CY4F, CY4H, CY4J, CY4K, CY4P, and CY4Q (see Figure 17). At the first five locations listed fluoranthene was detected, whereas at the last four locations both fluoranthene and benzo(b)fluoranthene were found. The detected signals from these contaminants, however, were fairly low and did not show distribution profile as clearly as the naphthalene class did (i.e., the contaminants were quite scattered). In addition, low concentrations of pyrene were detected at location CY4I.

Summed Fluorescence and Total Petroleum Hydrocarbon — A simplified presentation of the multichannel LIF data shows plots of total fluorescence detected in each excitation channel vs. depth. Similar results were also obtained for the field operation at Otis ANGB by adding together the fluorescence responses at all emission wavelengths for each excitation channel and normalizing them to the excitation photon intensity for the channel. The photon intensity was obtained from the averaged intensities measured before and after the push during the rhodamine B calibration routine. An example of such a plot is shown in Figure 33, from location CY4J, which was the most contaminated. Inspection of the total fluorescence vs. depth plots provides clear indication of the presence or absence of fluorescence signal for each channel. Comparison of the results for the different channels indicated qualitatively when the chemical nature of the contamination changed.

A "total sum fluorescence" response (for all excitation channels) vs. depth plot was also calculated in a similar manner. The total fluorescence response of all excitation channels and all emission channels for each push was obtained and normalized. The photon-normalized total summed fluorescence for each location that showed detectable LIF signals is shown in Figure 34. The error bars at the summed fluorescence figures are the result of both the error associated with the total fluorescence response, estimated by Poisson statistics, and the propagation of error in the calculations.

The soil sample data from Groundwater Analytical Laboratory indicated that the samples showed GC/FID characteristics that are similar to kerosene and petroleum products in the lubricating oil range (*n*-C18 to *n*-C36), with three- through six-ring PAHs. It is expected that there is a correlation between the summed fluorescence signals and the concentrations of TPHs as determined by the laboratory tests. The potential for providing CPT operators with the total summed fluorescence signal vs. depth in a real-time display during a sounding would be enhanced if a qualitative correlation with TPH could be demonstrated, as has been done with earlier-generation LIF-CPT instruments utilizing a single excitation channel.

Comparisons of laboratory results for TPH and total summed fluorescence for samples collected on-site are shown in Figure 35 (site CY4S) and Figure 36 (site CY4T). The comparison for in situ TPH detection at site CY4J and laboratory results is shown in Figures 37 and 38. Figure 37 gives the detail of contamination at the depth resolution for the measurements (2 cm for LIF and 2 ft for soil samples analyzed in laboratory), whereas Figure 38 shows the averaged measurements within the soil sampling depth (i.e., 2 ft). These plots indicate a rough correlation between TPH and total summed fluorescence, and the general trends of the two data sets appear to correspond. However,

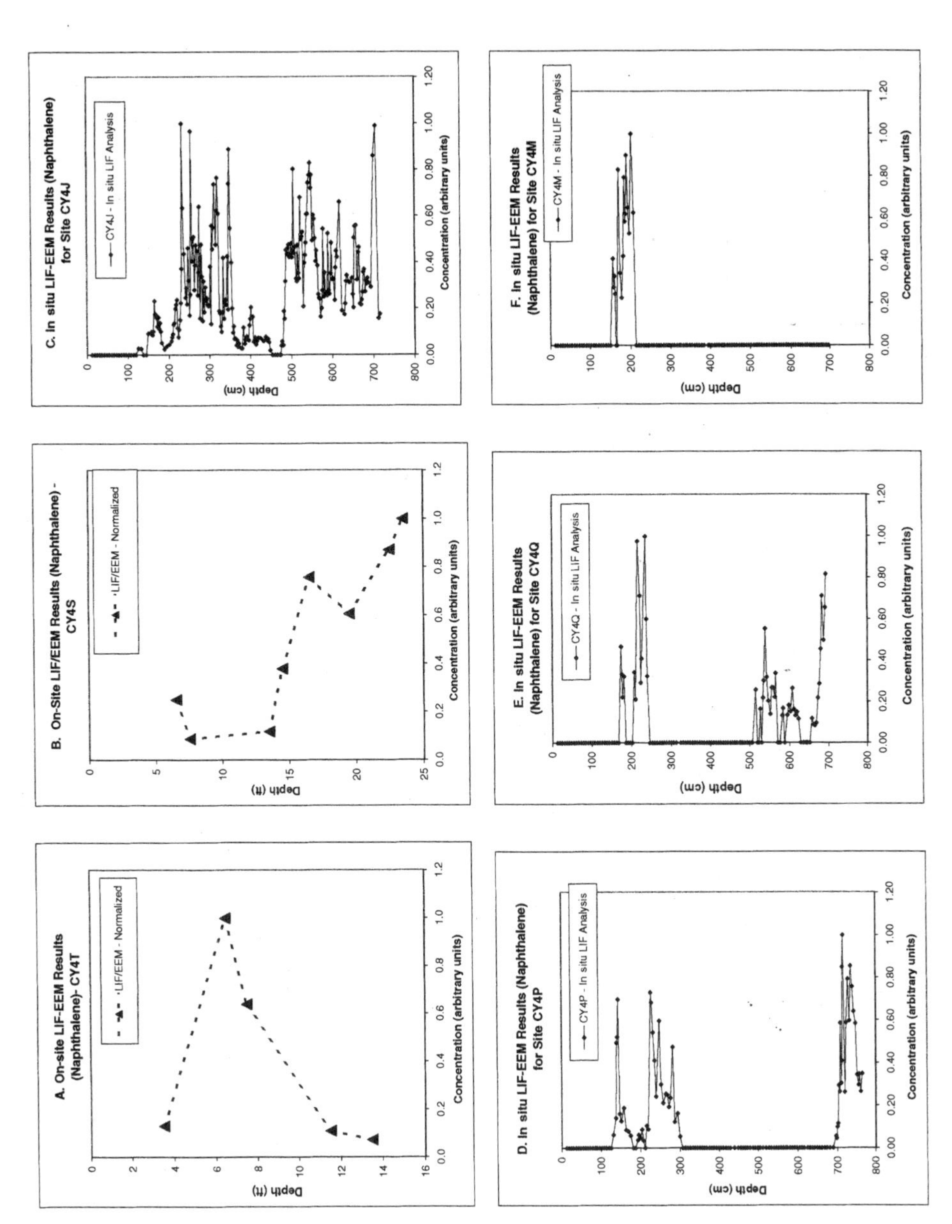

Figure 30 Concentration profiles of total naphthalene in various locations at Otis ANGB (relative scale).

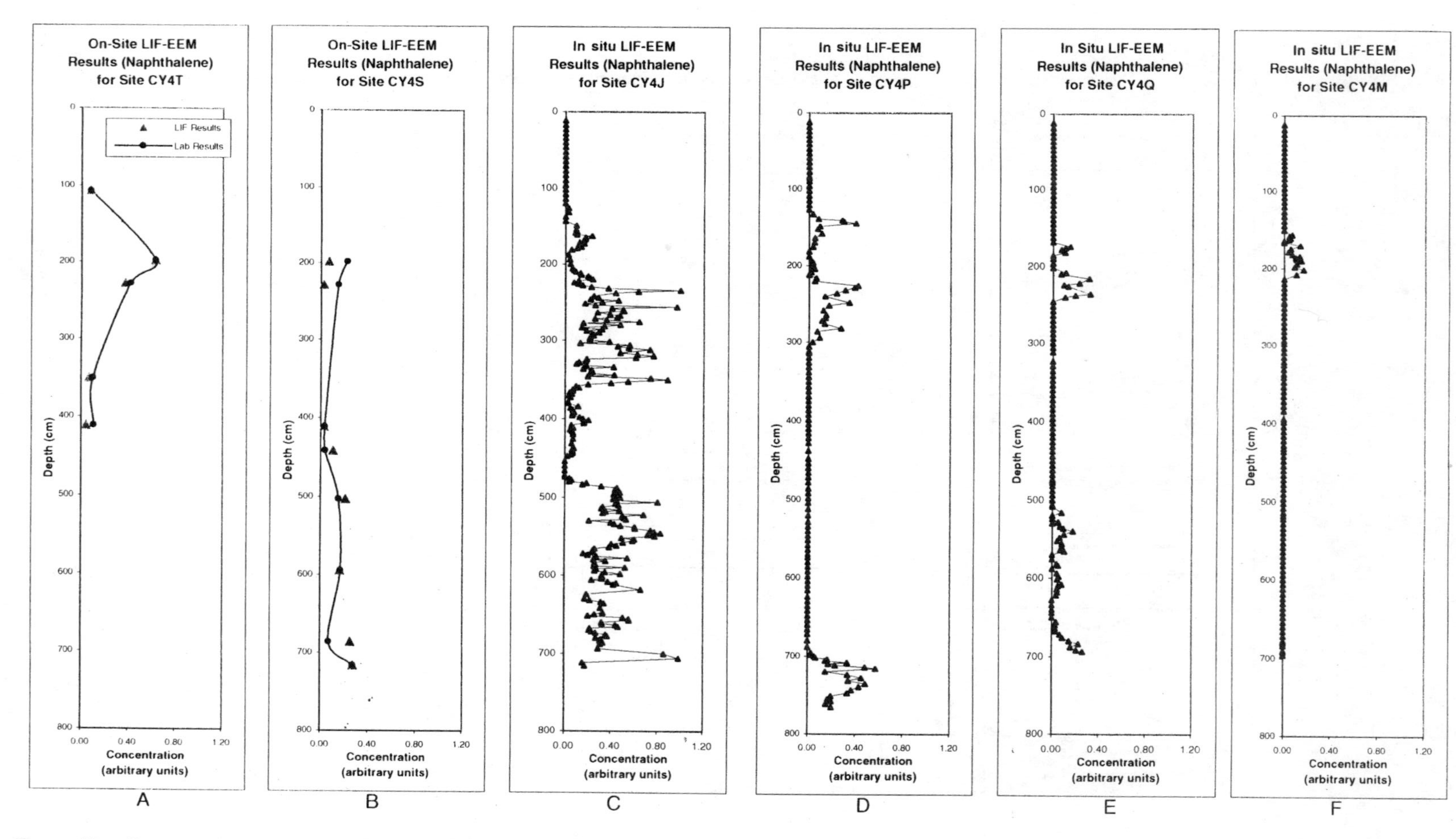

Figure 31 Concentration profiles of total naphthalene in various locations at Otis ANGB.

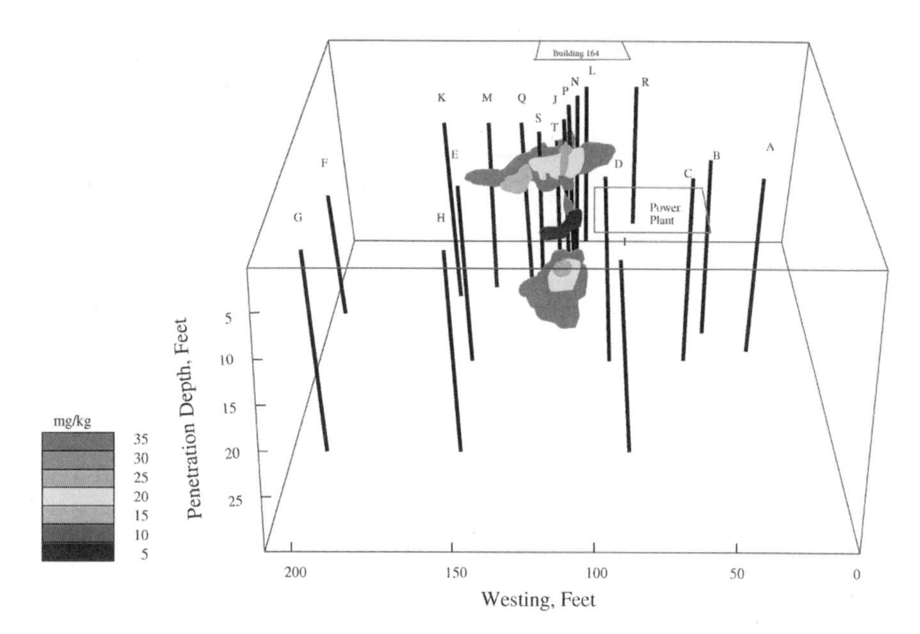

Figure 32 Distribution of total naphthalene (naphthalene and its derivatives) at Site CY-4 Otis ANGB.

the quantitative relationships between the data sets do not remain constant as a function of depth. This is not surprising because many nonfluorescent species will contribute to laboratory-measured TPH, and any quantitative correlation between TPH and total summed fluorescence would require the composition of the contaminant (both fluorescent and nonfluorescent components) to remain approximately constant everywhere at the site. Despite these limitations, a rough idea of the TPH distribution in the CY-4 site can be mapped by plotting total sum fluorescence (Figure 39).

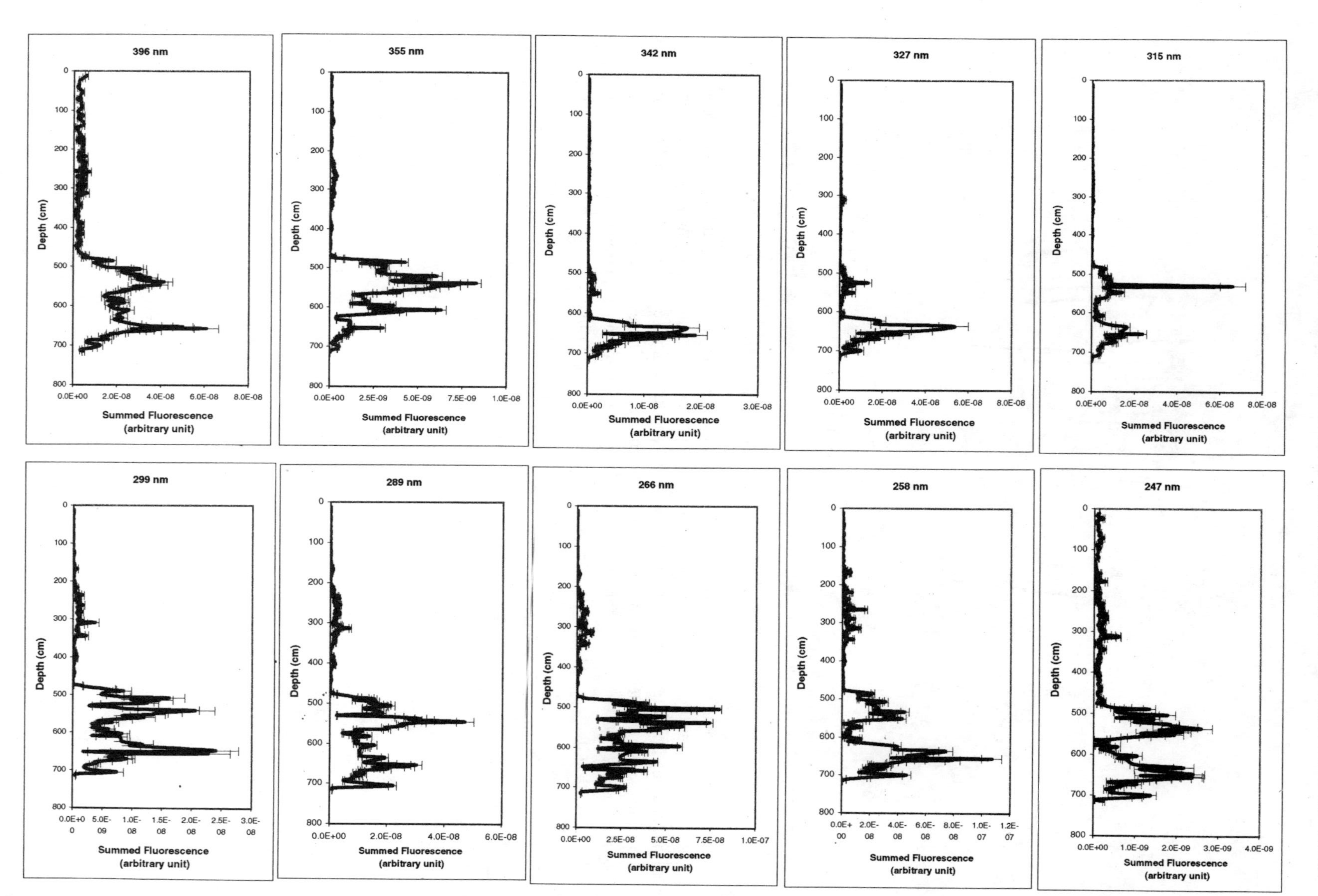

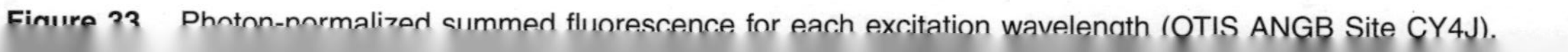

Figure 33 Photon-normalized summed fluorescence for each excitation wavelength (OTIS ANGB Site CY4J).

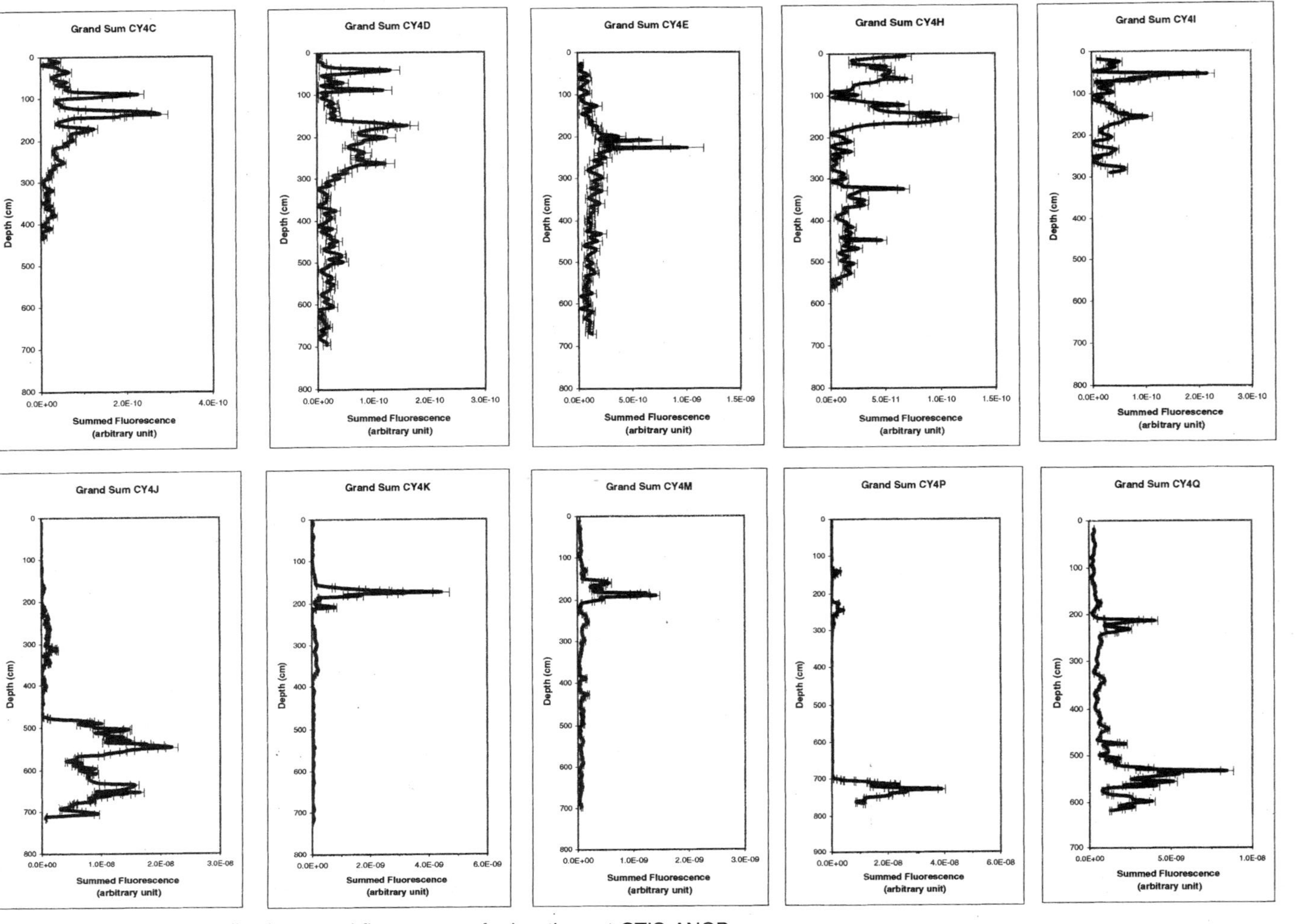

Figure 34 Photon-normalized summed fluorescence for locations at OTIS ANGB.

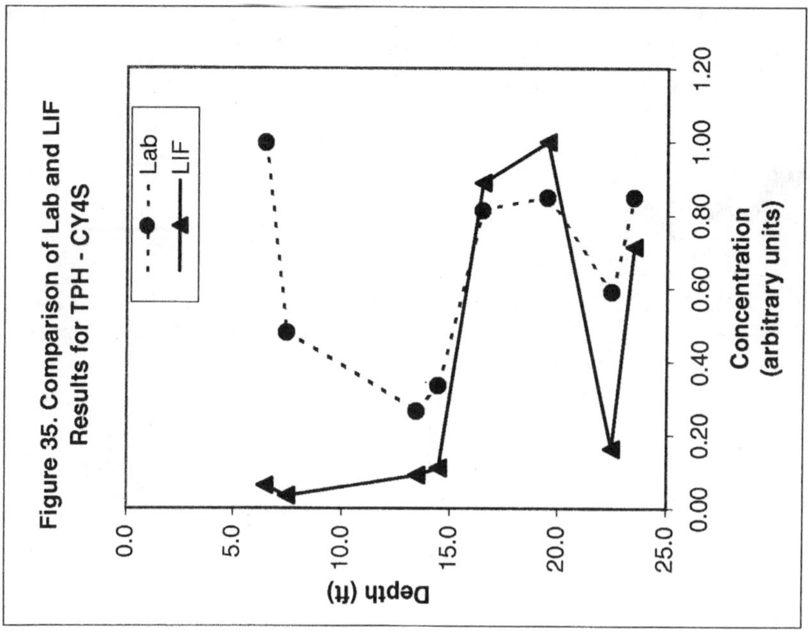

Figure 35. Comparison of Lab and LIF Results for TPH - CY4S
Lab
LIF
Depth (ft)
0.0
5.0
10.0
15.0
20.0
25.0
Concentration (arbitrary units)
0.00
0.20
0.40
0.60
0.80
1.00
1.20

Figure 35

Figure 36. Comparison of Lab and LIF Results for TPH - CY4T
Lab
LIF
Depth (ft)
0.0
2.0
4.0
6.0
8.0
10.0
12.0
14.0
16.0
Concentration (arbitrary units)
0.00
0.20
0.40
0.60
0.80
1.00
1.20

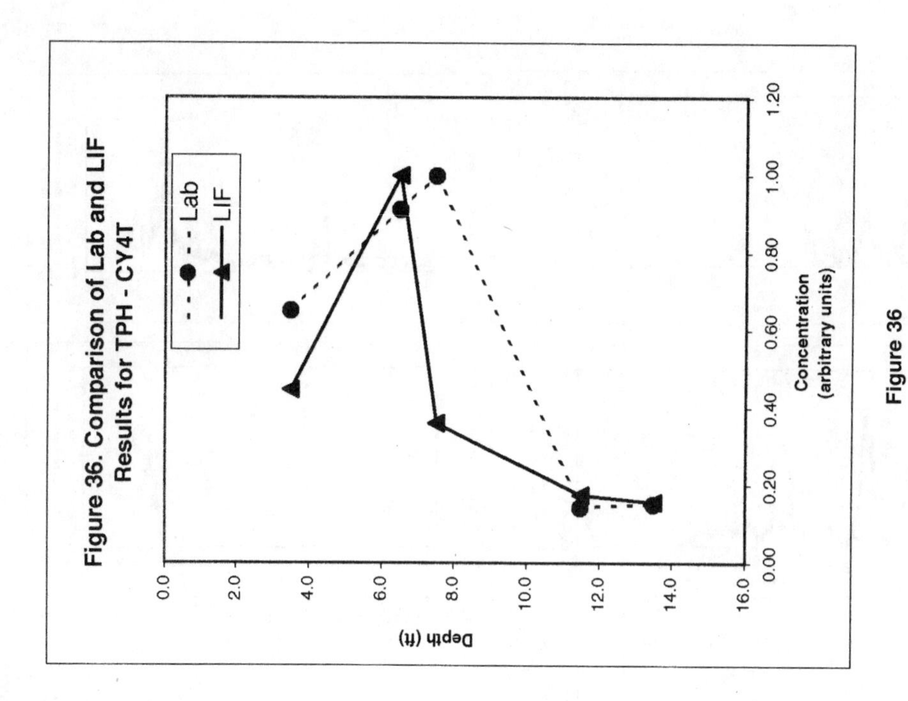

Figure 36

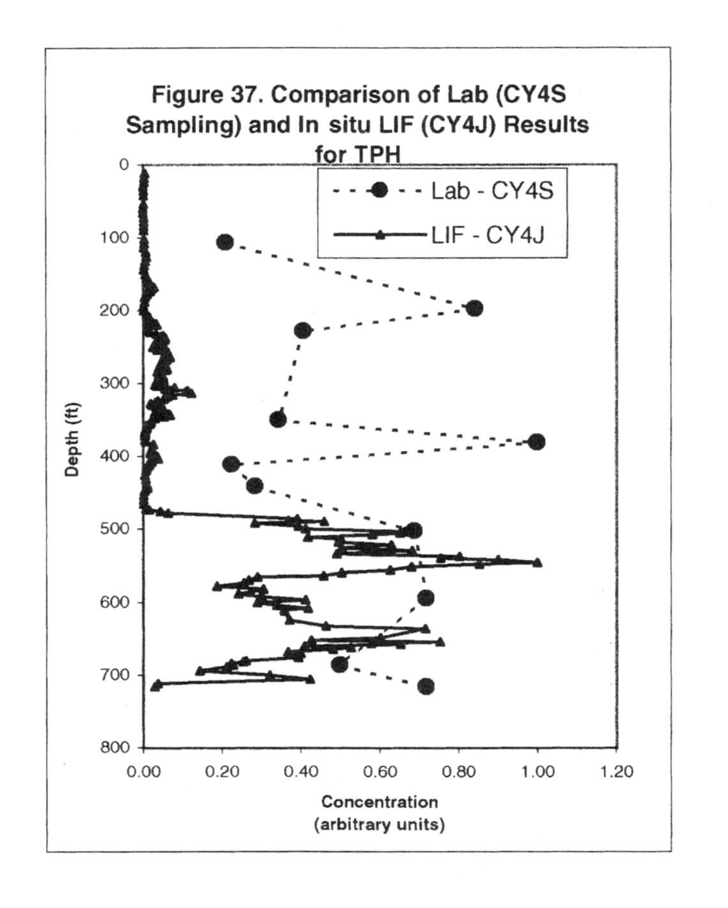

Figure 37

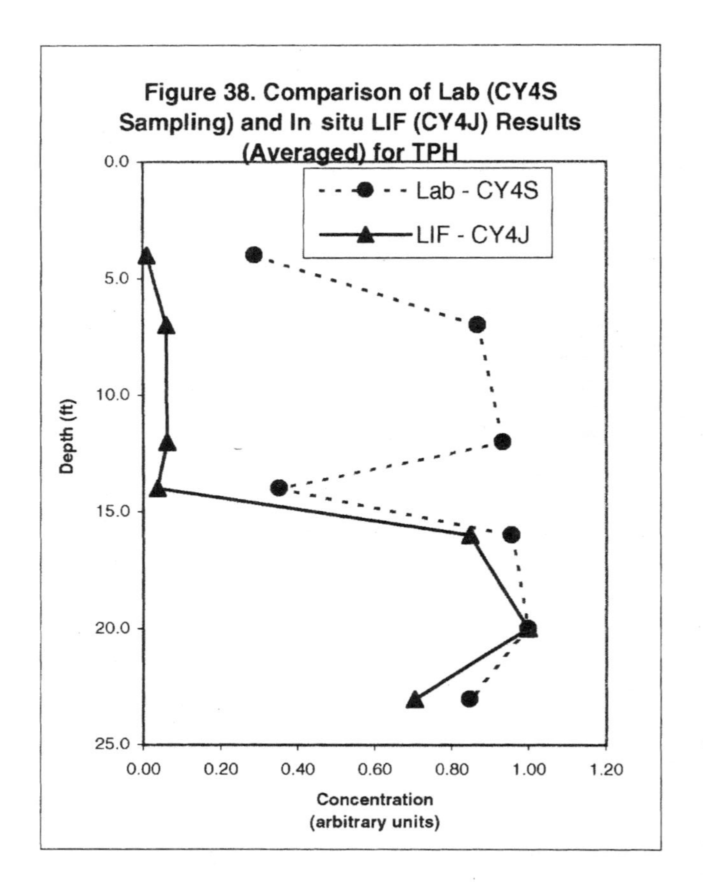

Figure 38

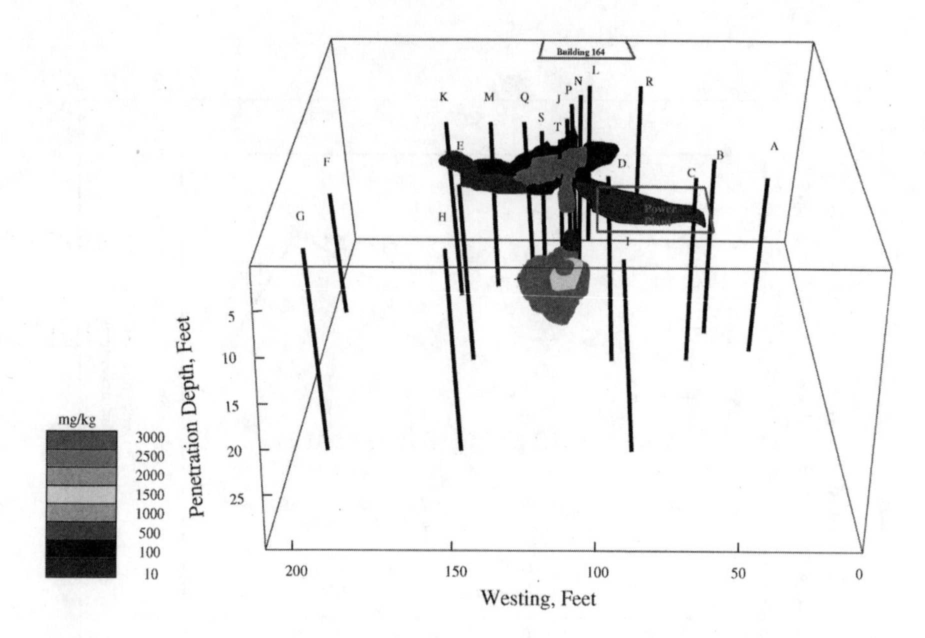

Figure 39 Distribution of total petroleum hydrocarbons (total summed fluorescence) at site CY-4 Otis ANGB.

CHAPTER 5

Outlook

5.1 DEVELOPMENTAL GOALS

Continuing developmental goals include the following:

- Improve the stability of the instrument through better management of heat and temperature in the laser module.
- Improve the laser intensity and hence the sensitivity through improvements in the optical system.
- Extend the list of target analytes to include more PAH and BTEX species and determine whether the technique can discriminate among them or sort them into classes.
- Demonstrate automated data collection and analysis.
- Incorporate real-time display of total summed fluorescence to assist operators in rapid site characterization.
- Extend field testing to sites with different soil characteristics.
- Provide a more extensive comparison of laboratory and in situ LIF results.

With the improvements outlined previously, the current instrument is expected to identify target analytes of polynuclear aromatic hydrocarbon (PAH) [and possibly benzene, toluene, ethylbenzene, and xylene (BTEX) at lower sensitivity] into classes containing one or more species and to determine concentration within a factor of 2 to 3 accuracy in soils similar to those encountered during this field work (sands and sandy silts). Laser-induced fluorescence (LIF) sensitivity for PAH should be equal to or better than conventional laboratory tests. Achieving both this accuracy and sensitivity in different settings depends on determining the effect of soil type (matrix) on the LIF detection and finding a suitable on-site calibration method to account for matrix effects.

5.2 POST-AATDF DEVELOPMENT

Instrument development has progressed in several aspects since the completion of the Advanced Applied Technology Development Facility (AATDF) project. A new water-cooling scheme has been developed for the laser operation, which efficiently dissipates heat generated by the laser power supply when it is operated in the cone penetrometer technology (CPT) truck. The cooling system gives the LIF equipment a higher degree of heat tolerance, which improves laser operation and stabilizes stimulated Raman scattering (SRS) output in the field.

A newer version of the SpectraMax™ software is also being evaluated, which includes a function to trigger data acquisition using an external source. This can be connected to the triggering output device on the CPT system and enable the data acquisition routine to be controlled internally. Such a triggering scheme would make LIF data acquisition fully automatic.

To reduce generation of scattered light during field work, a new probe has been designed and fabricated that consists of 12 small (2-mm diameter) sapphire windows embedded in a stainless steel metal button. This metal button replaces the sapphire window in the original design. The new design isolates each channel (excitation–emission pair) on the LIF probe and reduces light scattering and cross talk. The applicability of the new design is being evaluated in the field.

More extensive analysis has also been conducted on in situ LIF field data. The factor analysis methods have been further developed. Papers regarding the development of the methods and the results of the LIF data are in progress. Efforts are also being made to streamline these data analysis methods for application during in situ sampling.

5.3 FIELD TESTING

The system described previously has been modified to allow in situ measurement of Raman scattering from nonfluorescent samples using a near-infrared diode laser source. The sensitivity of this system[26] is comparable to that of a Raman system developed by Carrabba, which recently demonstrated its capability to detect dense nonaqueous phase liquid (DNAPL) at the U.S. Department of Energy (DOE) Savannah River site. A field demonstration has recently been completed[26] for the combined [first-generation U.S. Environmental Protection Agency (EPA)-funded] LIF and (AATDF-funded) Raman instrumentation at a former dry cleaner's site in Jacksonville, Florida. It was performed in conjunction with an EPA remediation project for DNAPL contamination.

During the initial site investigation, the multiwavelength fluorescence excitation probe detected nonfluorescent DNAPL, perchloroethylene (PCE), by the presence of dissolved fluorescent contaminants. Analysis of a sample extracted from the site before the field test indicated the presence of naphthalenes, *p*-xylene, and other, larger fluorescing aromatics, in the nonfluorescing PCE and provided the basis for applying LIF. In deploying the combination Raman (near-IR diode laser excitation) and LIF (two UV excitation wavelengths) probe at the site, the short-wavelength excitation in the two LIF channels produced fluorescence from dissolved PAHs in the PCE. The in situ fluorescence signals correlated with the depth distribution of DNAPL previously determined from soil borings. It was also noted that the long-wavelength Raman excitation source produced fluorescence of unknown molecular origin apparently unrelated to contaminant distribution, which precluded the observation of Raman signal from PCE and did not correlate with the known distribution of DNAPL.

These preliminary results indicate that the success of the Raman probe at Savannah River cannot necessarily be reproduced at other DNAPL sites. A combination of LIF and Raman approaches offer a better chance of delineating LNAPL and DNAPL plumes. The multiwavelength laser system offers a versatile tool for classification and semiquantitation of fluorescent contaminants and selective tracking of some nonfluorescent contaminants such as DNAPLs contaminated with fluorescent components.

CHAPTER 6

Applicability of the LIF-EEM Technology

The laser-induced fluorescence–excitation–emission matrix (LIF-EEM) probe advanced by cone penetrometer technology (CPT) equipment is an effective screening tool for site characterization and could provide an economical, rapid assessment of contaminated sites. Data resolution and the significant volume of subsurface information generated by this technique surpass conventional data collection methods and provide an excellent database to use with commercially available software for visualizing the three-dimensional extent of contaminants. This approach provides the technical information necessary to rapidly formulate a detailed conceptual model for remedial planning. The technology also has potential applications in monitoring various manufacturing processes and industrial wastewater operations.

Existing LIF systems are commercially available and include Fugro Geosciences' Rapid Optical Screening Tool (ROST™) and the U.S. Department of Defense Triservices SCAPS system. These systems have proven the applicability and effectiveness of LIF techniques to RCRA Facility Investigations, Corrective Measures Studies, and Remedial Investigations under a range of regulatory programs administered by state and federal agencies.

LIF data collected using these systems have been successfully applied to project objectives at industrial and commercial sites and government installations. Consequently, the LIF-EEM system is likely to be more readily accepted by regulatory agencies and end users.

6.1 DETECTABLE CONTAMINANTS

The LIF-EEM technique is applicable to all fluorescent contaminants and other fluorescent organic and inorganic materials. Many common contaminants such as gasoline, jet fuels, heating oil, diesel fuel, and coal tar as well as most other common petroleum products (e.g., petroleum oil lubricants) consist of mixtures of chemical compounds that include a significant fluorescent component. The presence of these materials allows detection of the contaminant mixture by the LIF-EEM technique. The range of detectable contaminants overlaps with the same common classes of contaminants of volatile and semivolatile organic compounds, TPH, and diesel range organics. It has been shown that, qualitatively, the technique can be used to efficiently detect PAHs according to their classes, and, quantitatively, it can determine the concentrations of contaminants to an accuracy of a factor of two for most of the cases. It provides much more information than some techniques that only report in "hit" and "nonhit" fashion.

Additionally, organic solvents may be detected by in situ fluorescence methods if they contain any fluorescent impurities. Many solvents released at industrial sites have comingled with fluorescent contaminants or acquired fluorescent organic and inorganic materials from soils when the solvents migrated. These fluorescent materials can serve as LIF markers for the organic solvents.

6.2 USE OF DATA FOR HUMAN HEALTH RISK ASSESSMENTS

The Total Petroleum Hydrocarbon Criteria Working Group (TPHCWG) established by the Texas Natural Resource Conservation Commission has developed a flexible approach for quantifying human health risk.[27] The approach uses aromatic fractions (e.g., carbon-chain lengths) to select a surrogate compound that has published toxicity factors during the assessment of human health risk. The LIF-EEM technique could be used to fractionate aromatic hydrocarbons along the lines of the TPHCWG approach.

With the use of excitation wavelengths at the lower end of the UV spectrum (247, 258, and 266 nm), the LIF-EEM instrument has the potential to be used for benzene, toulene, ethylbenzene, and xylene (BTEX) detection. The first generation of LIF system reported the detection of 3% benzene in cyclohexane solution. This percentage is similar to that of BTEX components in lighter jet fuels such as JP4. With the increased sensitivity of the current LIF system, significant signals (i.e., in several thousands of photon counts) have been obtained for 500 ppm benzene and 200 ppm *p*-xylene, both in cyclohexane solution. Fluorescence was still detectable (i.e., signal-to-noise ratio greater than 3) at 50 ppm *p*-xylene. A 300-ppm *p*-xylene cyclohexane solution was also used to saturate an Ottawa sand sample, and LIF signals were also detected. However, the latter situation is different from in situ soil measurements because the sample (i.e., *p*-xylene) was not fully mixed in sand and the higher wavelength excitation beams (i.e., 289 nm and up) were blocked to avoid the problem of scattered light interference.

At this stage, the LIF-EEM technology could be used as a qualitative and semiquantitative screening tool for site conditions. The ability to rapidly collect information to perform a semiquantitative human risk assessment will undoubtedly add to the value of this technology. It may be worth considering if the LIF technique could be used to fractionate aromatic hydrocarbons. This option would require additional soil sampling and analysis effort and in particular would require analysis of the BTEX constituents.

6.3 LIMITATIONS

Because CPT direct push equipment is the current mode of deployment for the probe, use of the instrument is dependent on the ability to penetrate site soils to the desired investigation depth. Consequently, the probe may not be useful in areas of crystalline rock, heavily cemented sediments, or locations with thick zones of coarse gravel or large cobbles. Under favorable, soft-soil conditions, the probe can be easily advanced to its current design limit of 50 ft.

In addition to soil texture, the depth of investigation by the LIF probe may be limited by detection limits that increase with depth (e.g., the probe will become less sensitive to contaminant concentrations). The multiple wavelengths of laser light used in the LIF-EEM analysis do not all traverse the optical fiber equally well, with attenuation of the shortest wavelengths tested to date. This results in higher detection limits for increased optical fiber length.

Although the EEM data analysis methods have been successful in analyses of fluorescent mixtures obtained in the laboratory and in the Otis field test, there is no guarantee of equal success in the case of field data for more complex mixtures. This is due to the limited number of laser wavelengths that reasonably can be launched into a fiber bundle, the generally unknown perturbation that the soil and other subsurface components can produce in the EEM of a given analyte, and the unknown number and type of reference EEMs needed. It appears that the technique is easily adapted to identification of contaminant classes but more difficult to fine tune for identification of individual chemical species. One challenge of this project and future related work is to better establish the applicability of these analysis techniques to field implementation in different soil types and with more complex mixtures of hydrocarbon contaminants.

6.4 EQUIPMENT DEVELOPMENT AND PATENT CONSIDERATIONS

The LIF-EEM instrumentation has significant potential, but additional technical and economic data will be required to achieve the commercialization goal. The Principal Investigator envisions two additional demonstrations of the technology over a 12- to 18-month period, assuming a suitable site can be found to demonstrate LIF-EEM capabilities. The objectives of the additional work are to demonstrate continuous operation on all channels, laser stability, automatic triggering and depth encoding of fluorescence, and near-real-time data processing capabilities. The results must be compared against analytical laboratory data to demonstrate acceptable correlations and to validate the chemical classification and quantitation achievable by the LIF-EEM method.

Research continues in an effort to collect information necessary to achieve the commercialization goal, which includes planned equipment demonstrations and sample analyses by an analytical laboratory.

The LIF-EEM technology has not yet been licensed. One patent is held on the technology by Tufts and a second patent is being pursued based on a modified probe design. Work in progress is likely to result in another patent application. Notably, up to three preexisting patents by others may need to be licensed in commercializing the LIF-EEM instrumentation.

CHAPTER 7

Comparison with other LIF Systems

Over the last several years, laser-induced fluorescence (LIF) technology has matured to the point where the equipment has transitioned to the private sector, and in situ LIF screening of petroleum hydrocarbons is now readily available as a commercial service. These sensors provide a method to obtain detailed information about the distribution of subsurface contamination prior to collecting soil samples or installing wells. LIF measurements continue to be most useful at screening a site for classes of hydrocarbon compounds rather than for identifying and quantifying individual contaminants.

Available systems include the U.S. Department of Defense Site Characterization and Analysis Penetrometer System (SCAPS),[29] which uses a single excitation wavelength LIF sensor, and Fugro Geosciences' Rapid Optical Screening Tool (ROST™),[30] which uses a tunable (selective excitation wavelength) dye laser. These in situ LIF systems have proven the applicability and effectiveness of LIF techniques during site characterization and remedial investigations. SCAPS is available for work on noncommercial government sites, and ROST™ is commercially available for work on government and private industry projects. A recent review of these and related technologies has appeared in the literature.[10,31]

This section provides a summary of engineering and operational features of the two commercially available systems and compares these features with the Tufts LIF system. More details of alternative configurations that have been deployed on occasion are available.[7]

7.1 COMMON COMPONENTS

Generally, the SCAPS, ROST™, and Tufts systems all use a platform consisting of a cone penetrometer technology (CPT) truck equipped with vertical hydraulic rams to advance an LIF probe into the subsurface. Each of the various probes is manufactured to deliver laser light to the soil by fiber optic cable through a sapphire window mounted near the probe's tip. The resulting fluorescence energy passes back through the window and travels through a second fiber optic cable to the surface for analysis. The optical fibers are integrated with the CPT geotechnical probe and electrical umbilical cable of the CPT system.

7.2 SCAPS SYSTEM[4,5,10,32,33]

The SCAPS LIF system was developed for in situ spectroscopic detection of petroleum hydrocarbons in soils. The system was developed in collaboration with the U.S. Army Engineer, Waterways Experiment Station, in a joint Army, Navy, and Air Force program.

The SCAPS LIF is a nonspecific field-screening technique that detects polycyclic aromatic hydrocarbon compounds with three and more aromatic rings. A key asset of this system is its simplicity; the laser has modest utility requirements and directly produces the excitation wavelength. The excitation fiber is mounted directly on the laser head, eliminating the need for an optical bench. Furthermore, the laser wavelength is efficiently transmitted by the optical fiber, so that depths as great as 50 m (or more) may be sampled. The greatest utility of the system is in rapid screening for petroleum, oil, and lubricant contamination to more precisely locate contaminated zones.

The primary limitation to the SCAPS LIF system is its inability to measure the dependence of fluorescence on additional parameters, which limits the equipment's ability to discriminate between different fluorescent contaminants. Further, the availability of only one excitation wavelength, 337 nm, makes many strongly fluorescent contaminants completely or nearly invisible to the instrument.

7.2.1 Engineering Design

The SCAPS system uses a pulsed nitrogen laser that operates at a wavelength of 337 nm as the light source, with a pulse energy of 1.4 mJ, and two silica optical fibers, one to conduct laser light to the sample and the second to convey fluorescence to the detector. The detection system consists of an intensified photodiode array (PDA) detector coupled to a spectrograph to quantify the fluorescence emission spectrum, typically providing a resolution of a few nanometers for the emission spectrum. Because the detector can be read quickly, it is possible to add spectra from multiple laser shots to improve the signal-to-noise ratio of the measurement. Typically, 10 laser shots are used per sample interval.

Generally, the PDA detector is activated only during the time period when the fluorescence signal is present. This approach reduces any contribution to the signal from background light and detector noise. Incrementing the delay of the detector gate for successive laser pulses also permits determination of fluorescence decay times, although this is not done in normal operation. Consequently, the fluorescence measurement performed by SCAPS is one-dimensional (e.g., fluorescence intensity is measured only as a function of emission wavelength). Studies have shown that differences in decay times are useful for discriminating compounds of environmental interest (e.g., polycyclic aromatic hydrocarbons) that cannot be resolved based on differences in their fluorescence emission spectra.

7.2.2 Data Presentation

Data acquisition is automated under computer software control. The computer sets and controls the sensor system and stores fluorescent emission spectra and strain gauge data. From the chemical fluorescence and soil characteristic measurements at each sampling depth, the SCAPS software extracts the maximum fluorescence intensity and associated peak wavelength and generates real-time raw fluorescence vs. depth plots and soil profiles.

Under normal operating conditions, fluorescence measurements are made at a rate of approximately one per second. For the standard CPT advancement rate of 2 cm per second, this corresponds to a vertical spatial resolution between measurements of 2 cm.

7.3 FUGRO GEOSCIENCES' ROST™ SYSTEM[6,7,10]

The ROST™ sensor evolved from the tunable laser instrumentation originally developed at North Dakota State University with U.S. Air Force research support. Following completion of a demonstration project at Tinker Air Force Base, the technology developers formed a small business, Dakota Industries, Inc., and participated in additional equipment demonstrations. The technology has been in commercial service to the environmental industry since 1994. The commercialization and initial marketing were performed by a government and industry consortium led by Loral Corporation and Dakota Technologies, Inc. ROST™ was acquired by Fugro Geosciences, Inc., in May 1996 and is now offered as an integrated service with their CPT systems worldwide.

An important advantage of this system over the SCAPS system is access to additional excitation wavelengths, which allows detection of light fuels and heavier molecular weight contaminants, and the addition of fluorescence decay time to the fluorescence measurement. ROST's ability to simultaneously collect and analyze three fluorescent parameters (intensity, emitted wavelengths, and fluorescence lifetime) without interruption of the test push allows continuous plotting of a contaminant's relative concentration and spectral signature with depth.

7.3.1 Engineering Design

The ROST™ sensor uses a wavelength-tunable ultraviolet laser coupled with an optical detector to measure fluorescence via optical fibers. Characteristic features of this system include a specially designed rhodamine dye laser, which is pumped by a primary neodymium-doped yttrium aluminum garnet (Nd:YAG) laser that produces 532-nm light at 50 hertz with a pulse energy of 50 mJ. The laser system is capable of generating wavelengths of light ranging from about 280 to 300 nm, depending on the dye being used. A 266-nm excitation wavelength capability is also available for detection of BTEX and other single-ring aromatic hydrocarbons.

The principal components of the ROST™ fluorescence detector consist of an Acton Spectropro 150, serving as a monochromator/spectrograph, a photomultiplier (PMT) tube, and digitial storage oscilloscope. The monochromator/spectrograph disperses the fluorescence emitted by the contaminants, which is then directed to an exit focal plane of the monchromator/spectrograph. Four optical fibers of progressively longer length are placed in the exit focal plane to collect the light and serve as delay lines to direct light of selected wavelengths to arrive at the PMT at selected time intervals.

The time delay feature allows the system to monitor emitted fluorescence at four wavelengths (340, 390, 440, and 490 nm) simultaneously during testing. These wavelengths were selected to allow detection of relatively light fuels (i.e., gasoline, jet fuel, diesel) and heavier molecular weight hydrocarbons (i.e., oils, coal tar, creosote) without the need to adjust monitoring wavelengths. Output from the PMT is directed to a digital storage oscilloscope, which digitizes, averages, and displays the fluorescence intensity vs. time/wavelength waveform. The waveform is then downloaded to the computer for permanent storage and postprocessing of the data.

7.3.2 Data Presentation

The CPT stratigraphic information and ROST™ contaminant screening data are displayed in real time on computer monitors as each test proceeds. CPT and ROST™ logs for each location are then printed in the CPT truck immediately following completion of each test sounding.

The ROST™ log presents a continuous plot of the fluorescence intensity with depth and the spectral signatures of detected petroleum hydrocarbons. Zones of affected soils can be identified by elevated fluorescence intensity and are differentiated from other potential petroleum products by the spectral signatures (waveforms) presented. An example ROST™ log is presented in Figure 40, which illustrates a typical fluorescence vs. depth log and waveform signatures of common petroleum hydrocarbon products. Waveform data are continuously collected and displayed during testing and can be compared to waveforms of common petroleum hydrocarbons or to bench-top waveforms of specific contaminants to identify and differentiate hydrocarbon compounds detected at a site.

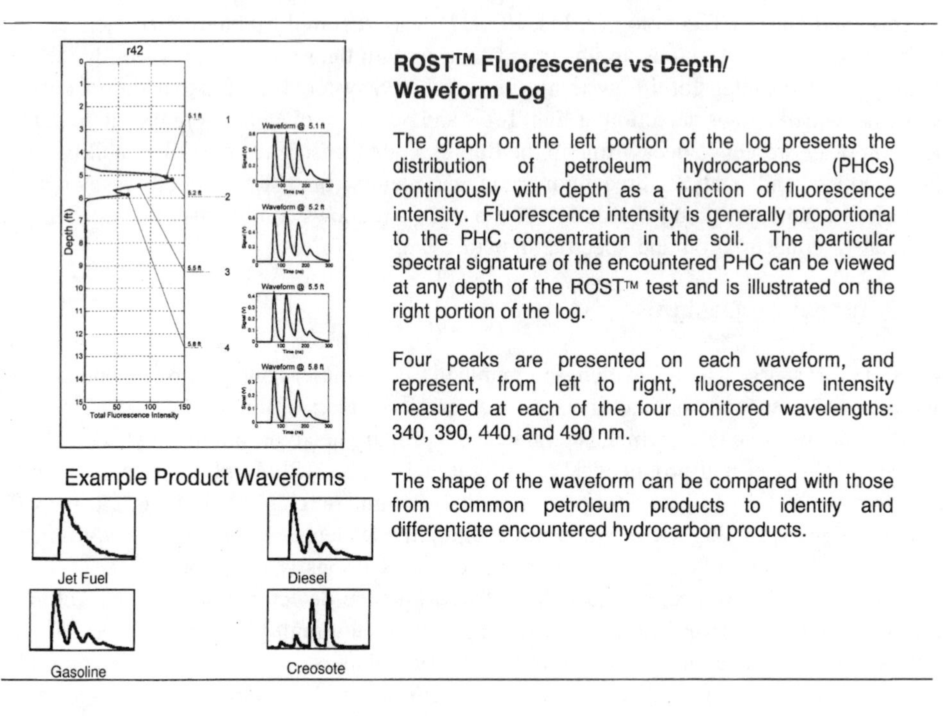

Figure 40

7.4 TUFTS UNIVERSITY SYSTEM

The Tufts LIF system advances the state-of-the-art by using multiple excitation wavelengths of laser light, and wavelength analysis of contaminant fluorescence, to form a three-dimensional fingerprint of the contaminant. The major advantages of this system are the wide range of excitation wavelengths available, which allows for excitation of most fluorescent contaminants at one of their strongest absorption bands, the simultaneous collection of both dimensions of fluorescence data, and simplicity of the instrumentation. Multiple-excitation wavelength screening provides for fewer contaminants to remain undetected. Detailed information on the engineering design of this system was presented earlier in this report.

Further points of comparison between the Tufts and ROST™ systems will be summarized below. These systems are not directly comparable to the SCAPS system because at the time of this study SCAPS did not generally collect three-dimensional data.

- The ROST™ system uses circulating laser dyes to generate excitation wavelengths of light. The Tufts system uses a Raman shifter.

- To obtain a three-dimensional fingerprint of the petroleum-hydrocarbon contaminant being screened, the Tufts and ROST™ systems both collect fluorescence intensity as a function of emission wavelength. The Tufts system also uses excitation wavelength as a second dimension, whereas ROST™ uses fluorescence decay time. An important difference between the two approaches is that the excitation spectrum should generally contain more information than the temporal decay curve.
- Both systems display the total fluorescence intensity continuously with depth. However, at the current stage of LIF software development, Tufts researchers are still working toward continuous collection and display of spectral contaminant signatures, whereas ROST™ already has the capability to continuously collect and display this information during testing. The Tufts LIF data management system currently requires the probe to be stopped to collect spectral signatures.

7.5 TYPICAL COSTS

The principal cost advantages of these LIF systems, when compared to conventional drilling and sampling techniques, are associated with the reduction in field investigation time, the number of soil borings and monitor wells, and the quantity of samples collected from a site. During a general site characterization effort, LIF techniques can provide more data in less time and less expensively than a conventional drilling and sampling approach. Importantly, worker exposure to contaminants is reduced using these in situ technologies.

ROST™ is available worldwide for a cost of approximately $4000 to 4500 per day or site-specific footage rates, which includes a CPT rig. Crew per diem and equipment mobilization are additional and site specific. Typical crew members include a ROST™ system operator, CPT operator, and an assistant. Under normal conditions, an average of 300 ft can be completed in a day at, depending on project specifics, roughly $10 to $15 per foot.

Cost data were not readily available on the SCAPS system, which is typically mobilized for site characterization work at Department of Defense installations. Cost data were also not directly available for the Tufts system because additional technical and economic data will be required to achieve the commercialization goal for this sensor. However, a minimum cost estimate was calculated for purposes of comparison. It was assumed that the cost associated with instrument usage would be its retail value, $150,000 (approximately twice the retail component costs), recovered over a 3-year period with 260 working days per year for a daily instrument cost of $192, adjusted to $300 per day for maintenance and repairs. With no further design development, the minimum cost per day for using the Tufts instrumentation was calculated at $2300 per day for the CPT vehicle and crew, $1000 per day for the additional LIF crew member, plus $300 per day for the instrumentation. This provides a minimum estimated daily charge of $3600.

References

1. M. Sauer, J. Arden-Jacob, K. H. Drexhage, F. Gobel, U. Lieberwirth, K. Muhlegger, R. Muller, J. Wolfrum, C. Zander. 1998. "Time-resolved identification of individual mononucleotide molecules in aqueous solution with pulsed semiconductor lasers," *Bioimaging*, 6(1), 14–24.
2. M. Sauer, K. H. Drexhage, C. Zander, J. Wolfrum. 1996. "Diode laser based detection of single molecules in solutions," *Chem. Phys. Lett.*, 254(3-4), 223–228.
3. B. B. Haab, R. A. Mathies. 1995. "Single-molecule fluorescence burst detection of DNA fragments separated capillary electrophoresis," *Anal. Chem.*, 67(18), 3253–3260.
4. S. H. Lieberman, G. A. Theriault, S. S. Cooper, P. G. Malone, R. S. Olsen, P. W. Lurk. 1991. In *Second International Symposium on Field Screening Methods for Hazardous Wastes and Toxic Chemicals*, Air & Waste Management Association, Pittsburgh, PA, 57–63.
5. D. S. Knowles, S. H. Lieberman, M. Davey, K. E. Stokeley. 1997. In Proceedings, *Field Analytical Methods for Hazardous Wastes and Toxic Chemicals*, Air & Waste Management Association, Pittsburgh, PA, 147–158.
6. R. W. St. Germain, G. D. Gillispie. 1991. "Transportable tunable dye laser for field analysis of aromatic hydrocarbons in groundwater," *Proc. Second International Symposium of Field Screening Methods for Hazardous Waste Site Investigations*, Las Vegas, NV, Feb. 11–13.
7. G. D. Gillispie, R. W. St. Germain. 1995. In Proceedings, *Field Analytical Methods for Hazardous Wastes and Toxic Chemicals*, Air & Waste Management Association, Pittsburgh, PA, 478–489.
8. A. Henderson-Kinney, J. E. Kenny. 1995. "Spectroscopy in the field: emerging techniques for on-site environmental measurements," *Spectroscopy*, 10 (7), 32.
9. S. J. Hart, Y.-M. Chen, J. E. Kenny, B. K. Lien, T. W. Best. 1997. "Field demonstration of a multichannel fiber optic laser induced fluorescence system in a cone penetrometer vehicle," *Field Anal. Chem. Technol.*, 1(6), 343–355.
10. S. H. Lieberman. 1998. "Direct-push, fluorescence-based sensor systems for in situ measurement of petroleum hydrocarbons in soils," *Field Anal. Chem. Technol.*, *2*, 63–73.
11. J. Lin, S. J. Hart, T. A. Taylor, J. E. Kenny. 1994. "Laser fluorescence EEM probe for cone penetrometer pollution analysis," *Proc. SPIE*, 2367, 70.
12. J. Lin, S. J. Hart, W. Wang, D. Namychkin, J. E. Kenny. 1995. "Subsurface contaminant monitoring by laser fluorescence excitation-emission spectroscopy in a cone penetrometer probe," *Proc. SPIE*, *2504*, 59.
13. S. J. Hart, Y.-M. Chen, B. K. Lien, J. E. Kenny. 1996. "A fiber optic multichannel laser spectrometer system for remote fluorescence detection in soils," *Proc. SPIE*, 2835, 73.
14. S. J. Hart. 1998. Ph.D. thesis, Tufts University, Medford, MA.
15. S. Mathew. 1997. Ph.D. thesis, Tufts University, Medford, MA.

16a. I. M. Warner, E. R. Davidson, G. D. Christian. "Quantitative analyses of multicomponent fluorescence data by the methods of least squares and non-negative least sum of errors," 1977. *Anal. Chem.*, 49, 2155.

16b. I. M. Warner, G. D. Christian, E. R. Davidson, J. B. Collis. "Analysis of multicomponent fluorescence data," 1977. *Anal. Chem.*, 49, 564.

17. G. B. Jarvis, S. Mathew, J. E. Kenny. 1994. "Evaluation of Nd:YAG-pumped Raman shifter as a broad-spectrum light source," *Appl. Optics*, 33, 4938.
18. A. O. Wright, J. W. Pepper, J. E. Kenny. 1999. "Fluorescence measurement using an angled two-fiber probe," submitted to *Anal. Chem.*.
19. J. W. Pepper, A. O. Wright, J. E. Kenny. 1999. "In situ measurements of subsurface contaminants with a multichannel laser-induced fluorescence system," submitted to *Environ. Sci. Technol.*

20. J. Lin, S. J. Hart, J. E. Kenny. 1996. "Improved two-fiber probe for in situ spectroscopic measurements," *Anal. Chem.*, 68, 3098.
21. E. R. Malinowski. 1991. *Factor Analysis in Chemistry*, 2nd ed. John Wiley & Sons: New York.
22. A. Lorber. 1984. *Anal. Chim. Acta*, 164, 293.
23. A. Lorber. 1985. *Anal. Chem.*, 57, 2395.
24. E. Sanchez, E. R. Kowalski. 1986. *Anal. Chem.*, 58, 496.
25. S. H. Lieberman, D. S. Knowles, W. C. McGinnis, M. Davey, P. M. Stang, D. McHugh. 1995. In Proceedings, *4th International Symposium-Field Screening Methods for Hazardous Waste and Toxic Chemicals,* Las Vegas, NV, 22–24 February.
26. K. Warner, W. D. R. Premasiri, G. J. Hall, S. J. Hart, J. W. Pepper, J. E. Kenny. (submitted to) "Comparison of LIF and Raman spectroscopies for detection of subsurface DNAPL using cone penetrometer probes: a case study," *Environ. Sci. Technol.*
27. W. H. Weisman. 1988. "A risk-based approach for the management of total petroleum hydrocarbons in soil," *J. Soil Contam.*, 7(1), 1–15.
28. Y.-M. Chen, J. W. Pepper, J. E. Kenny. 1999. "Identifying and quantifying chemical composition from real-world EEM data using factor analysis method," submitted to *Anal. Chem.*
29. "Site Characterization and Analysis Penetrometer System (SCAPS) for Rapid Site Characterization and Monitoring," Workshop 7, 10th Annual Conference on Contaminated Soils, University of Massachusetts at Amherst, October 24, 1995.
30. U. S. Environmental Protection Agency. 1997. "The Rapid Optical Screening Tool (ROST™) Laser-Induced Fluorescence (LIF) system for screening of petroleum hydrocarbons in subsurface soils," document EPA/600/R-97/020, February.
31. A. Henderson-Kinney, J. E. Kenny. 1995. "Spectroscopy; overview of selected field technologies," CFAST Report (September 1995), 33–38.
32. Naval Command, Control, and Ocean Surveillance Center. 1992. "Subsurface screening of petroleum hydrocarbons in soils via laser induced fluorometry over optical fibers with a cone penetrometer system," S. H. Lieberman, S. E. Apitz, L. M. Borbridge, and G. A. Theriault, NCCOC, RDT&E Division. Paper presented at the International Symposium on Environmental Sensing, Berlin, Germany, June 1992.
33. G. A. Theriault, R. Newberry, J. M. Andrews, S. E. Apitz, S. H. Lieberman. 1992. "Fiber optic fluorometer based on a dual wavelength laser excitation source," paper presented at O/E Fibers 1992, September, Boston, Massachusetts.

APPENDIX A

Design Manual for the Laser-Induced Fluorescence Cone Penetrometer Tool

Jonathan E. Kenny, Jane W. Pepper, and Andrew O. Wright

A.1 INTRODUCTION

This design manual describes the construction of a laser-induced fluorescence (LIF) sensing tool for use with cone penetrometry (CPT) remote sensing technology in soils. LIF is a common spectroscopic measurement technique that offers high sensitivity and selectivity toward those compounds that fluoresce. In particular, polyaromatic hydrocarbons (PAHs), which are notorious trace constituents of hydrocarbon fuels and which are known to have a hazardous health and environmental impact, can be detected by means of LIF.

Conventional LIF techniques are often directed at collection of total emission, with no distinction among different compounds that are emitting light. For in situ analysis of soils using CPT, total LIF emission from hydrocarbon mixtures containing PAHs has served as a useful method for rapid detection and monitoring of contaminant plumes arising from surface chemical or fuel spills.

However, mixtures can be characterized with greater detail because each component exhibits unique wavelength dependence in excitation and fluorescence profiles. This provides a fingerprint that can be used to identify the contribution of a compound to a total mixture emission.

The approach outlined in this manual is directed at generating excitation and emission wavelength-dependent fluorescence intensities of contaminant mixtures instantaneously during the course of a CPT push. Such real-time measurements are achieved by using multiple channel excitation and detection. Multiple laser wavelengths are generated simultaneously by means of stimulated Raman scattering (SRS) and charge coupled device (CCD) detector arrays allowing multichannel detection. Excitation–emission matrices (EEMs) that result provide the potential for not only determining total contaminant plumes but also for distinguishing particular compounds or classes of compounds within the plumes without sacrificing the inherent speed and efficiency of CPT technology.

Researchers at Tufts University have performed LIF-EEM-CPT assessment of ground contamination at Hanscom U.S. Air Force Base (AFB) in Massachusetts and at U.S. Coast Guard Station in Elizabeth City, North Carolina, with an instrument developed in association with the U.S. Environmental Protection Agency (EPA). The instrumentation described herein represents a second-generation LIF-EEM-CPT tool funded by the AATDF with the end goal of commercialization. It utilizes a number of design changes implemented since assembly of the EPA tool.

The four principal components of the LIF tool, as in many spectroscopic sensing systems, are the excitation system including the laser and Raman gain medium, the optical delivery system of fibers, the sample interface on the multichannel probe, and the detection system on the CCD/Spectrograph.

A.2 EXPERIMENTAL STRATEGY

The overall goal is to produce a commercially viable LIF-CPT instrument that allows detailed and accurate on-site assessment of PAH contamination in soils. This will improve on the capacity provided by commercial LIF-CPT technology such as SCAPS (Tri-Service Characterization and Analysis Penetrometer System), which provides contaminant mapping with minimal distinction of chemical content and ROST (Rapid Optical Screening Tool), which requires penetrometer stoppage to acquire its chemically specific information.

Using the first-generation EPA tool developed at Tufts,[1,2] excitation emission matrices (EEMs) for chemical characterization have been acquired more quickly than with ROST. However, the EPA tool still required stoppage, and its CPT probe used multiple windows over an extended depth interval so that EEMs had to be depth-corrected in postacquisition analysis.

Further improvements still sought, beyond present capabilities of SCAPS, ROST, and the first EPA tool, include the following:

- Extend the overall range of excitation wavelengths available to better separate PAH responses.
- Reduce the area of the sampling region required at the probe interface.
- Extend the overall sensitivity of the multiple-channel technique through efficient light collection and reduction of instrumental artifacts.
- Provide sensitive detection of benzene, toluene, ethylbenzene, and xylene compounds.
- Streamline data processing and user interface during data collection.
- Allow real time and automatic operation without having to interrupt a CPT push for specific data collection.
- Reduce the instrumental space and utilities necessary for deployment in a CPT vehicle.

A.3 THEORY OF COMPONENTS

A.3.1 Properties of Lasers and Stimulated Raman Scattering (SRS)

Lasers have been increasingly used as excitation sources over conventional lamps due to lasers' unique and favorable light propagation characteristics. In addition to being very monochromatic, laser output is far more intense, propagates in a defined path, and exhibits coherent phase relationships that enhance many physical and material responses, such as SRS.

As an example of the advantage presented by lasers, consider a mercury (Hg) lamp that has a specified emission of 50 W (Oriel Corporation). At peak emission (about 365 nm) the lamp exhibits close to 100 $mWm^{-2}nm^{-1}$ at a distance of 0.5 m. A high-speed lens system (e.g., f/0.7) will convert this to 14 $mWnm^{-1}$ of collimated output. On the other hand, a standard calibration helium-neon (HeNe) laser with 0.5 mW output and a bandwidth of 0.004 nm yields 125 $mWnm^{-1}$, nearly one order of magnitude improvement in usable output power delivered with two orders of magnitude reduction in total power generated.

Lasers have two types of output configuration: continuous wave (cw) and pulsed. The HeNe laser in the above example is a cw laser, where a steady output is maintained over time. A pulsed laser's output, however, is concentrated into short bursts. A typical laser pulse width is between 10^{-6} and 10^{-9} seconds, and pulsed lasers usually operate at repetition rates of less than 1000 Hz.

In LIF experiments, where a laser acts as the excitation source, cw lasers may allow better sensitivity than pulsed lasers because the latter exhibit low *average* power due to their duty cycle. However, pulsed lasers have extremely high *peak* powers over the duration of a pulse, making them figure prominently in LIF strategies in a number of other different ways. Perhaps most importantly, important nonlinear processes affecting wavelength of light (e.g., frequency doublers and optical parametric oscillators) are more easily implemented using the high peak powers of pulsed laser systems. Even conventional dye laser systems demonstrate greater ease of design and operation in a pulsed system. As a result, a single pulsed system (such as a Nd:YAG laser) is capable of generating virtually all wavelengths in the visible and UV ranges.

The high peak powers associated with pulsed laser systems also generate stimulated Raman scattering. Raman scattering is a process by which incoming laser light is coupled with a molecule's vibrational energy (or energies) in such a manner that a new field is generated, having an energy that is the sum or difference of the vibrational energy and the original energy of the photon. Normally this is an incoherent process, but if the laser light intensity is sufficient and the molecules are in some sort of optical cavity coherent buildup of stimulated emission occurs and the Raman-shifted light begins to act as a laser also. Furthermore, higher order vibrational shifts (second, third, etc.) become similarly stimulated as the input laser intensity increases. Because the action associated with the Raman process is rapid, SRS can occur even within a nanosecond pulse, and SRS output is similarly pulsed. The generation of SRS is a matter of attaining a minimum threshold of light energy for the process to occur, explaining why peak power is the important feature for the laser that generates (or "pumps") the process.

The Raman medium for SRS is typically a pressurized gas in a long pathlength stainless steel cell, with transparent windows on either end to allow light throughput and contain the gas. The choice of gas is usually predicated on simple vibrational structure. H_2 and CH_4 gas are typical SRS gain media and have been used successfully for years. H_2/CH_4 mixed gas systems have been developed and characterized at Tufts University that offer numerous lines from the two vibrational energies arising from CH_4 and H_2 (at 2916 and 4155 cm^{-1}, respectively) and their higher-order contributions (overtones and combinations of the frequencies).[3] All resulting SRS lines at different wavelengths are generated simultaneously, thereby providing a significant advantage over other methods of generating altered wavelengths whose output must be scanned over time.

The SRS wavelengths are colinear and must be separated by a dispersing system, such as a prism or grating system. After separation, the wavelengths can be captured independently as they follow different paths (Figure A1).

A.3.2 Light Throughput in Optical Fibers

Having been a mainstay within the telecommunications industry for decades, optical fibers are finding increasing application in opto-electric device technology and scientific instrumentation. Optical fibers are commonly used for remote LIF sensing and for LIF-CPT in particular. Although most LIF-CPT applications have employed single excitation and detection fibers, the AATDF application employs up to 24 such fibers, amounting to 12 simultaneous LIF measurements.

The principle on which use of optical fibers is based is *total internal reflection*. Consider light traveling through the core of the fiber with the core material having a relatively high refractive index. Around the core is a cladding layer having a slightly smaller refractive index. As light travels through the core, it reflects off the boundary with the cladding layer. According to basic principles of optics, if the angle of incidence with the boundary is sufficiently large with respect to surface normal (i.e., greater than the "critical angle," θ_c), then total reflection occurs with no loss of energy. As a result, optical fiber has good transmission efficiency for light traveling at high-incidence angles to its circumferential boundaries or at small angles with respect to the fiber axis.

The optical fiber described above is classified as a "step index" fiber, where there is a discrete boundary between the core and cladding materials that have different refractive indices. There is

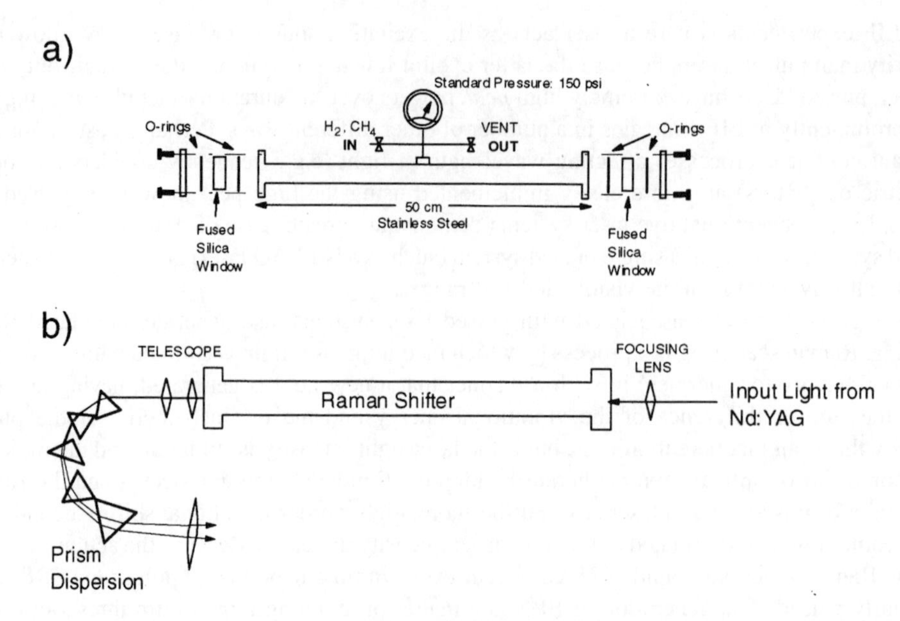

Figure A1. Basic design of Raman cell (a) and representation of SRS and dispersion in LIF-EEM experiment (b). Note that only two out of many output wavelengths are shown.

also a "graded index" configuration. where light is bent through material having a continuously varying refractive index. Furthermore, depending on the core diameter, optical fibers can be single mode or multiple mode, where modes represent different paths for light spatially distributed in the fiber during transmission. Single-mode fibers are used in many laser applications where a uniform spatial profile is necessary. In these LIF experiments, however, high spatial integrity is not necessary, and the larger size of the multimode fibers provides significant advantages for ease of optimizing system throughput.

The simple structure of a step-index fiber includes a core material through which light travels, a cladding material of lower refractive index around the core, and a buffer layer of polymer for mechanical strength and flexibility. Typically, the core and cladding materials are the same, with slightly different impurities to alter the refractive indicies.

An important characteristic of an optical fiber is its numerical aperture (NA), defined as $\eta\mathbf{sin}(\theta_c)$, where η is the refractive index of the medium being traversed by light. When light comes out of a fiber, it spreads out according to the numerical aperture. Conversely, the ability of a fiber to collect light is also dependent on numerical aperture. When designing any optical system, including optical fiber systems, NA is important and must be accounted for to accurately model performance (Figure A2).

To collect incoming laser light with an optical fiber end, the beam divergence of the laser is matched with the NA of the fiber. To attain this optimum situation, lenses are carefully selected and placed at discrete distances from one another and the optical fiber end. Additionally, the laser beam diameter must also be smaller than the fiber core diameter for optimum coupling, which can also be addressed by lens selection and placement.

Different materials can be used to construct optical fibers. Important material characteristics include strength, flexibility, and transparency in the wavelength region being transmitted. Fibers are most commonly used in near-infrared applications, and the preferred material for this application is flexible glass or even plastic. The key property required for the LIF-CPT tool described, however, is transparency in the UV region, which requires that the fiber be constructed of high-grade fused silica. Fused silica is brittle, so it is usually coated with a flexible buffer layer of polyimide to reduce stresses from bending and twisting the fiber.

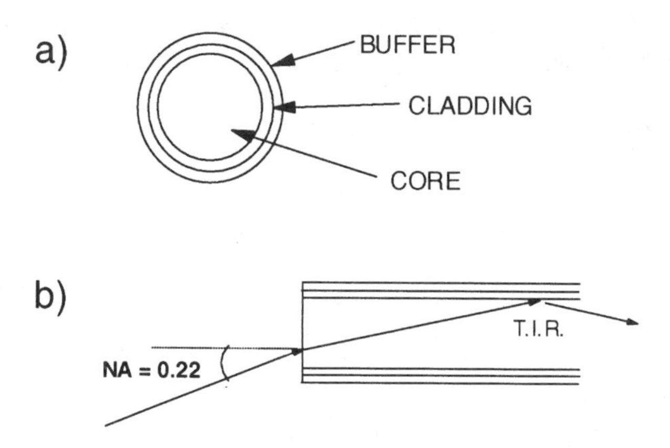

Figure A2. Layout of step-index multimode fiber for light transmission. (a) End-on view with structure. (b) Side view, with numerical aperture and total internal reflection demonstrated.

Although fused silica is used for UV light transmission, it is not truly transparent. The standard high-grade material exhibits an increasing attenuation (or loss) below 400 nm. Such loss is scaled in dB/m, so the longer the length of fiber the more significant the attenuation. If one considers a 10-m length of fiber, the percent transmission is ~89% at 400 nm, ~71% at 300 nm, and ~45% at 250 nm; however, for a 20-m length the respective transmissions are ~79%, ~50%, and ~20%. A further unavoidable limitation of UV light transmission through fibers is a process called "solarization," where absorbing defects are created within the fiber due to passage of the UV radiation. Solarization represents a gradual reduction in optical fiber throughput and is additive to the natural attenuation associated with the optical fiber (Figure A3).

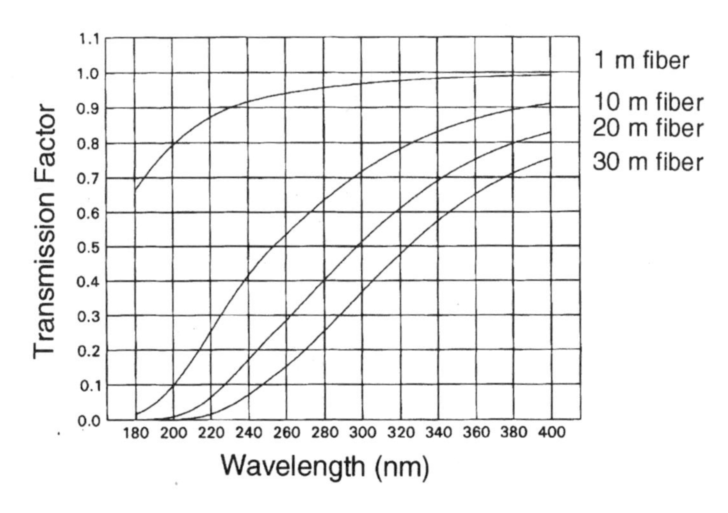

Figure A3. Transmission of UV-transmitting fused silica fiber (Polymicro) as a function of wavelength and fiber length.

A final limitation with optical fiber delivery of laser excitation light is the occurrence of laser-induced damage. For transparent materials such as optical fibers, damage thresholds typically approach 1 GW/cm^2. This peak power level is approached when 6 mJ of laser energy, over a 4-nsec pulse, is focused down to the core diameter of a 400-μm fiber. Such conditions are entirely within reach for this LIF-CPT system. Typically, optical materials have a specified bulk damage threshold and surface damage threshold, where the latter is generally lower. For optical fibers, however, damage is even more likely to occur at the core-cladding interface, where the first internal reflections occur.[4] This type of damage has been seen and required moving fiber ends away from optimum coupling with laser light near the laser's focus.

A.3.3 Probe Interface

The specific demands on the sampling interface, i.e., window in LIF-CPT, are unique. Primarily, it has to be rigid enough to withstand the sample pressures and harsh treatment associated with CPT work. Furthermore, it has to allow continuous sample acquisition so as to not restrict the total number of samples that can be required in a single push, and be reusable from push to push.

For the instrument described here, all of the excitation/detection channels have been integrated into a single window. Previous versions of this LIF-EEM-CPT instrument employed a separate window for each excitation wavelength, arrayed vertically along a CPT pipe segment.

The probe interface needs a transparent window to allow light passage to and from fiber ends, which are too fragile to be in direct contact with soil. The window material itself must be strong enough to withstand soil contact and is most often composed of sapphire. The geometry of the excitation and detection fibers must be optimized to account for the 2-mm thickness of the sapphire window, so the surface of the expanded excitation light in a channel is completely overlapped with the surface that can be collected by the detection fiber.

Finally, probe design must integrate the unique requirement that many fibers be used. Use of 10 channels at 2 fibers per channel would require 20 fibers that must be integrated into the probe. Economical bundling and termination of fibers must be considered in the design to contain the probe within an existing CPT rod diameter.

A.3.4 Optical Detection System

A final key component of any spectroscopic system is detection of optical response of an analyte that is excited with incoming light. Upon excitation, a number of different optical responses occur, such as scattering (e.g., Rayleigh and Raman scattering) and luminescence. The principles of optical detection are generally the same in either case.

A.3.4.1 Charge-Coupled Devices vs. Photomultipler Tubes

The choice of detector is predicated on the wavelength regime of light being monitored, strength of the light, required temporal response to light, and dynamic range between full signal and zero signal cases. Most detectors are solid-state electronic devices, where current or potential is measured as a response from a photoactive element that captures light. For LIF-CPT, light detection occurs principally in the visible and ultraviolet portions of the spectrum. For this regime, the two most common types of detectors are photomultiplier tubes (PMTs) and semiconductor arrays. In the semiconductor category, photodiode arrays (PDAs) have been extensively used until recently when more powerful charge-coupled devices (CCDs) were introduced. The system used in the LIF-CPT tool is a CCD detector.

PMT and CCD detection systems offer different advantages. Briefly, a PMT has greatest advantage when detection of a transient response is required. This type of detector provides a fast response that is directly monitored by a current, is inexpensive, and shows favorable conversion efficiencies for light to current ("quantum efficiency"). The PMT is particularly useful for temporal domain spectroscopies and photon counting, and it can generally be considered the workhorse for much of the available optical spectroscopic detection instrumentation. A further advantage associated with a PMT is its large dynamic range between low light levels and high light levels and its easily modified response by adjusting the power voltage. Where PMT detectors begin to suffer a disadvantage is when multiple wavelengths of light are being collected out of a spectrograph. An array detector can collect discrete wavelength channels simultaneously, whereas a PMT can only measure wavelengths one at a time via scanning. In essence, an array consists of many small detectors on a one- or two-dimensional surface all operating simultaneously in an integrated system.

CCD detectors offer further advantages to the multichannel advantage described previously. Previously, PDA detectors had low efficiencies and were one-dimensional linear arrays. CCDs, however, offer higher quantum efficiencies and lower noise than even PMTs and are the most sensitive detectors currently available. Furthermore, CCD detectors allow two-dimensional imaging, which increases the amount of information that can be acquired in real time.

For generating LIF-EEMs in soil under low light conditions with two-dimensional information processing, the choice of a CCD over a PMT detector is clear.

A.3.4.2 CCD Operation and Binning

CCDs are valuable for spectroscopic applications due to their low noise and their ability to segregate individual pixels into user-defined configurations, in a process called "binning." A CCD detector consists of a two-dimensional array of pixels, arranged into rows and columns. For each pixel, a photon of light generates a charge that is maintained in that element until all the pixels are read to an output amplifier for data processing. Registers are used along the rows and columns of the CCD so that each pixel is processed individually by moving charge packets sequentially closer to the output amplifier. First, the column elements of the first row are sequentially processed, then the elements of the second row, and so forth. The noise associated with this charge transfer process is negligible and represents the readout noise of the amplifier.

Binning occurs by combining charge packets of pixels in the readout process. The entire detector surface could be defined as one "superpixel," where it acts as a single photoactive element, or as smaller superpixel elements, depending on the application. For imaging purposes, binning is usually not desired because this limits resolution of the readout image. For spectroscopic purposes, it is common to bin all the column pixels together for each row to provide intensity vs. wavelength information. In multichannel applications such as LIF-EEM, column pixels are binned according to where fluorescence from each channel is being imaged along the vertical dimension of the CCD.

A.3.4.3 Spectrograph Selection and Wavelength Response

Wavelength information from a fluorescence signal is processed along the horizontal axis of the CCD. For this to occur, the CCD must be placed at the output (image) end of an imaging spectrograph. A spectrograph typically consists of an entrance slit aperture, a collimating mirror, a diffraction grating, a refocusing mirror, and an exit port on which the detector is placed. Spectrographs differ from spectrometers in that the spectrometer employs slits at the output aperture to select a single wavelength of throughput light.

Important criteria of a spectrograph are the numerical aperture, the effective focal length of the collimating and refocusing mirrors, and the dispersion (either angular dispersion or linear reciprocal dispersion). The aperture defines the optimum light propagation that neither underfills nor overfills surfaces of the mirrors and grating. Overfilling reduces light throughput, whereas underfilling lowers wavelength resolution. The focal length also affects wavelength resolution, with larger spectrographs providing better-resolved wavelengths.

Dispersion is a property of the spectrograph grating. The grating's groove density affects its ability to disperse light, which is mathematically related to the angular dispersion $d\theta/d\lambda$. A high angular dispersion means a relatively small change in wavelength is efficiently dispersed. Reciprocal linear dispersion is often more useful because it incorporates the spectrograph focal length and is defined as $d\lambda/dx$, where dx corresponds to a spatial separation on the output image plane of the spectrograph. Using reciprocal linear dispersion, it is easy to calculate the $d\lambda$ resolution element per discrete pixel size (dx) of a CCD detector.

An important consideration in the detection system, for both the spectrograph and CCD, is the wavelength response. The spectrograph's response is based on its grating and described as the "blaze angle" of its grooves. The blaze angle provides a wavelength of maximum reflection response, typically 50 to 70%. To the low and high end of this maximum, the response tapers away from the maximum, and it is necessary to specify a blaze wavelength optimized for the wavelength region of light being detected.

The CCD response is a material property that does not vary greatly from device to device; however, modifications to improve the UV response of CCDs should be mentioned. One UV response improvement is "back thinning," which reduces the effective thickness of the photoactive surface to avoid attenuation of low-wavelength light within the surface prior to creation of a photoresponse. The other modification is an antireflective coating on the CCD to enhance UV and visible response (Figure A4).

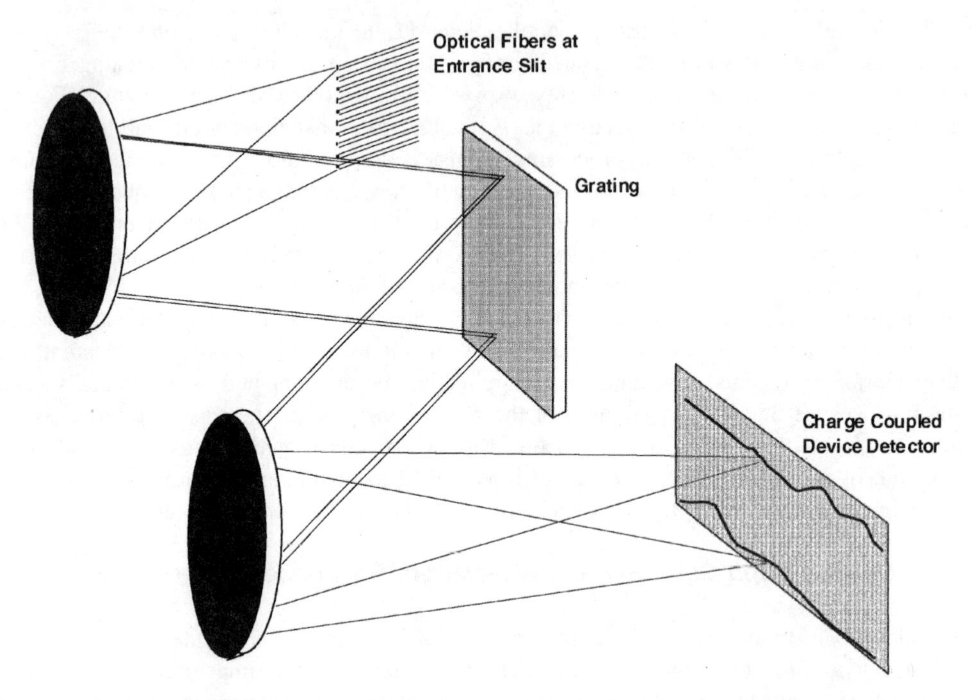

Figure A4. Representation of multichannel LIF-EEM data acquisition through spectrograph elements and CCD detector.

A.3.4.4 Light throughput and Sensitivity in CCD/Spectrograph Systems

In spectroscopic measurement, light levels being detected are often either excessive or very weak. With high levels, saturation (nonlinear electrical response) is a problem, whereas with low light levels the signal response might not exceed the intrinsic noise of the detector. In either case, it is necessary to adjust the light throughput of the detection system.

Spectrograph light levels are controlled through adjustment of entrance slits, which are essential for establishing the appropriate aperture for light collection and subsequent refocusing at the exit plane. Light being detected is usually focused onto the slit aperture, and slits will typically block some portion of this light. Therefore, it is common with low light levels to open the slits wide at the expense of wavelength resolution. Conversely, the slit width can be narrowed to regain resolution when light throughput is high.

With CCD detection, the amount of time the pixels are exposed to light determines the sum response that is obtained. Thus, the integration time of the CCD can be lowered or raised depending on the amount of light flux being measured. At very low levels, integration times of hours are feasible, although stray light must be rigorously controlled in this case. With high light levels, integration times must be kept sufficiently short to avoid "blooming," which occurs when a pixel or superpixel has too many photo-generated charges resulting in spillover to adjacent pixels/superpixels.

A.3.4.5 Image Quality in CCD/Spectrograph Systems

For imaging experiments, such as LIF-EEM acquisition with a CCD, data quality may be influenced by image aberrations from optics inside the spectrograph. Typical aberrations include edge distortions and depth-of-focus distortions. Modern spectrographs use optics that minimize certain other aberrations, such as "toroidal" as opposed to "spherical" optical systems, which contain inherent aberrations. Appropriate imaging is essential to assure that spectral information is uniformly processed across the surface of the CCD detector. Image blurring associated with poor depth of focus can lead to overlap and corruption of information, which is one of the causes of "cross talk" at the detector interface, and its effect must be minimized.

Blooming, discussed previously, is also a main source of cross talk and must also be minimized accordingly. In the course of this manual, two sources of cross talk are mentioned, including detector interface cross talk (discussed previously) and sample interface cross talk (at the LIF-CPT probe, described subsequently). These are separate issues, and it is helpful to keep track of which source of cross talk is being considered.

A.3.5 Cone Penetrometry

Fugro Geosciences CPT testing equipment was used for direct push delivery of the LIF probe during field work. CPT includes a selection of tools that perform different functions when hydraulically advanced through soils. The tools are deployed through the floor of a specially designed truck that is self-contained and climate controlled and houses operating controls, electrical and optical cables, computers, and ancillary equipment.

The CPT truck weighs approximately 20 to 25 tons and has high ground clearance for negotiating uneven or brushy terrain present at various testing locations. The outside dimensions of the truck are approximately 30-ft long, 8.5-ft wide, and 13-ft high. The vehicle generates AC power (110 V 60 cycle) for lighting and computer equipment.

The CPT truck uses a hydraulic ram mounted in the forward portion of the truck, near the operator area, to push 1-m long threaded steel rods into the ground. These rods advance the cone at a constant rate of 2-cm s (ASTM Standard D 3441-86) and convey the CPT electrical and LIF optical cables from the downhole instruments to on-board computers. Approximately 400 ft of rods can be stored on the truck in the push-rod rack.

The CPT truck is accompanied by a support truck, which pulls a trailer containing a high-pressure steam cleaner, grout mixer, and pump. The steam cleaner is used to clean downhole equipment before each sounding. The grout mixer is used to prepare grout and cement mixtures that are tremied into CPT soundings following LIF collection. The support truck carries a supply of potable water for cleaning and grouting purposes.

Particular advantages of using the CPT system for advancement of the LIF probe are

- CPT equipment can advance the LIF probe at a controlled rate and can recognize and record subtle changes in soil conditions in the same interval the LIF probe collects fluorescence data.
- LIF data collection systems are housed inside the cone truck, allowing work to proceed in adverse weather without affecting data quality.
- The cone penetrometer direct push approach reduces potential exposure of site personnel to contaminants.

- There are no soil cuttings generated during LIF equipment advancement and, consequently, minimal materials to dispose of as potentially hazardous.

Stratigraphic data are acquired with the electric cone penetrometer contained within the cone tip. This electric cone tip is used to advance the LIF probe and obtain continuous stratigraphic data. It consists of a cylinder-shaped device housed in stainless steel and threaded at its upper end to receive the steel rods by which it is pushed. The cone contains two resistance strain-gauge load cells for independent measurement of resistance at the cone tip and sleeve friction as the unit is advanced through soil. Soil classifications are performed using the cone tip resistance and sleeve friction ratio.

A.4 DESIGN AND ASSEMBLY OF COMPONENTS

A.4.1 Laser Excitation Table

The laser excitation table consists of the following parts: optical breadboard and cover, laser, Raman shifter and optics, prism dispersion system, and multichannel optical fiber launch system (Figure A5).

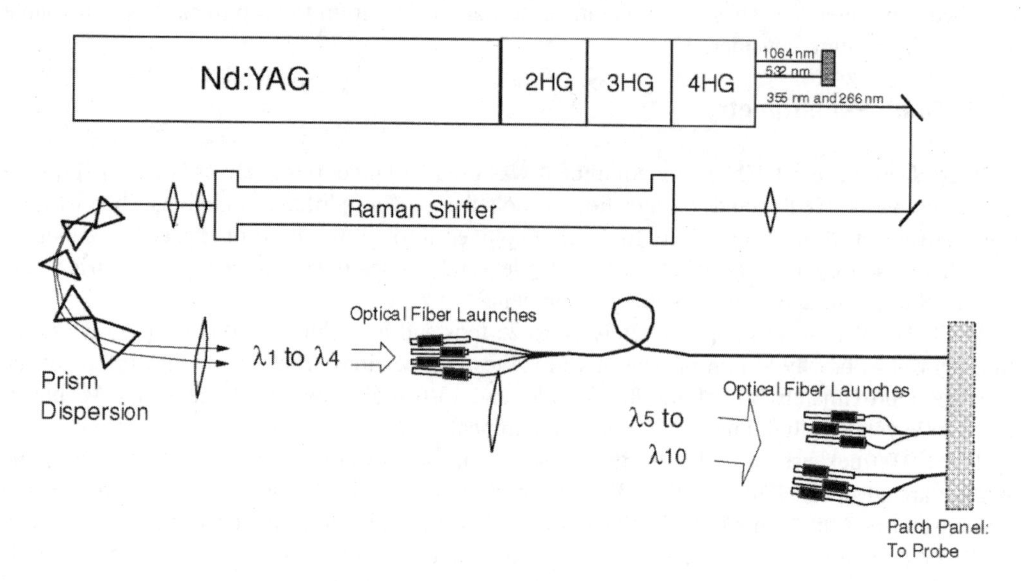

Figure A5. Layout of the optical table used in the current LIF-EEM instrument.

The *optical breadboard* is a standard platform in spectroscopic applications and serves two purposes. It provides base support and vibration isolation and a means for convenient and rigid placement of equipment and optical components. The placement of equipment is facilitated by a periodic array of screw holes on the breadboard surface. The breadboard selected for this LIF-CPT application is a standard item manufactured by Technical Manufacturing Corporation (TMC). The Tufts lab had successful experience with this manufacturer when assembling the EPA experimental system. The board is a 75 Series lightweight breadboard (75SSC-119-02), 2 in thick with a table area 23 × 47 in and $^1/_4$ in holes on 1 in centers and weighing 74 lb. The dimensions represent a significant reduction in the size and weight of the laser excitation table, from the first generation EPA instrument, allowing easier transport and installation in the CPT vehicle without sacrifice of rigidity or vibration dampening capacity. The cover for the breadboard is designed by Case

Technologies, Inc., and placed directly around the perimeter of the board. It is lightweight and has a lid that is shut during normal operation and allows convenient access during experimental adjustment and troubleshooting.

The *laser* is a pulsed flashlamp-pumped Nd:YAG operating at 50 Hz, which provides simultaneous and collinear output of 354.7- and 266.0-nm wavelengths. The model "Brilliant" is manufactured by Quantel S.A. in France.

There were three basic motivations in the selection of this laser:

- Small size and durability
- Simultaneous output of 354.7 and 266.0 nm
- Duty cycle of the pulsed output

The size of the laser is relatively small for a system of its output characteristics, and more importantly it can operate off a standard 120V/30A power source. The laser consists of a power supply and the head and harmonic modules that generate the laser emission. The power supply rests on the floor, and the head and harmonic modules sit on the optical table surface with the other components of the excitation system. The power supply and laser emission components are connected by an umbilical. The system is cooled during operation by a closed-cycle water to air heat exchanger that can deliver approximately 1 kW to the ambient surroundings. The small size and power utility requirements as well as ease of replacement of the flash lamps that have a limited lifetime make it an appropriate field operable laser.

The delivery of two wavelengths is a unique feature for Nd:YAG laser technology and products. The dual-wavelength delivery is accomplished by the successive staging of the second, third, and fourth harmonic modules. In the fourth harmonic module, a pair of dichroic mirrors direct 266.0- and 354.7-nm (fourth and third harmonics) emission to the output and 532- and 1064-nm (second harmonic and fundamental) emissions to a beam dump. The multiple wavelength separation feature at the dichroic mirrors is a nonstandard approach offered by Quantel prior to the selection of the laser and represents a novel development in laser applications. The fabrication of the mirrors progressed through numerous stages before a coating was obtained that could withstand the multiple-wavelength requirements without damage.

The duty cycle of the laser output was also a principal consideration for laser selection. The repetition rate of 50 Hz for the Quantel Brilliant is an improvement over more conventional 10- and 20-Hz systems due to the higher average power produced for a given pulse energy. Furthermore, its specifications excel over other 50 Hz systems in its short pulse width that yields higher peak power for a given pulse energy. The operating specifications of 6 mJ for each of the 266.0- and 354.7-nm wavelengths, with pulse duration under 6 nsec, is adequate for attaining SRS gain thresholds. The laser is mounted on a platform fabricated of aluminum with the optical height of laser output at 5 inches.

With the *Raman shifter*, a standard 1-in cylindrical gas cell, 50 cm long, is used. The main body was fabricated out of stainless steel. Fused silica windows, 1-in diameter and 3/8-in thickness provided by Esco Products, Inc., are mounted on each side into o-ring seal flanges that contain the operating gas at pressures up to 250 psi used during experimentation and field work. Valves and a gas regulator are also attached to the shifter to monitor pressures and to provide easy filling and flushing of gas. The Raman shifter is filled with H_2 and CH_4 gas to 150 psi under normal conditions.

The shifter is placed into two v-groove mounts produced by Oriel Corporation, which in turn are mounted on a precision optical rail (New Focus, Inc.) so that focusing and collimated lenses can be adjusted along an aligned axis. Fused-silica turning prisms are used to direct the laser beams into the Raman shifter. Iris diaphragms help to define the light path at either end of the optical rail, and a 50-cm focal length lens focuses light within the body of the shifter. A 10 mm/100 mm lens pair forms a telescope for recollimating the laser beams and SRS to a beam diameter of approximately 1 in. Accessory optics for the Raman shifter are fabricated of fused silica and mounted on stands produced by Newport Corporation and New Focus, Inc. Stands are fixed in place using specialized optical rail mounts or onto the breadboard surface by means of tie down clamps (Newport Corporation).

Multiple wavelengths are generated by SRS. This feature combined with the fact that two different laser wavelengths are pumping the system result in the 10 wavelengths used for excitation of PAHs in soils, which are listed in Table 1.

Table 1. SRS Wavelengths Used for LIF-CPT Probe

Wavelength (nm)	SRS Peak Assignment (*F, H, M*)
246.9	4, 0, 1
257.5	4, 1, –1
266.0	4, 0, 0
288.4	4, 0, –1
299.1	4, –1, 0
314.8	4, 0, –2
327.6	4, –1, –1
341.5	4, –2, 0
354.7	3, 0, 0
416.0	3 ,–1, 0

In this tabulation, F relates to the fundamental beam (4 for fourth harmonic output at 266.0 nm, 3 for third harmonic output at 354.7 nm), H is the number of hydrogen vibrational quanta associated with the output, and M is the number of methane vibrational quanta. Negative values for H and M correspond to Stokes processes and positive values to anti-Stokes. Other wavelengths arise from additional combination processes and are generally weaker than wavelengths in Table 1; however, those wavelengths that might be used include 239.5 nm (4, 1, 0), 275.1 nm (4, –1, 1), 309.1 nm (3, 1, 0), 321.4 nm (3, 0, 1), and 395.6 nm (3, 0, –1).

Five *prisms* in the LIF instrumentation are made of fused silica and produced by Esco Products, Inc. They are placed on an aluminum table surface mounted on posts (New Focus, Inc.), so the prism centers are 5 in off the table surface. Double-sided adhesive tape keeps the prisms in position, although a more durable mounting can be devised as needed. Angled jigs fabricated at the Tufts University machine shop provide convenient placement of the prisms with respect to one another (Figure A6).

A 300-mm fused-silica lens provides focusing for the first stage of multichannel *optical fiber launches*. Two 150-mm fused-silica lenses are placed side by side to capture remaining dispersed SRS lines and refocus them into the second stage of launches. Selection of the focusing optics was discussed in the preceding section.

Each launch assembly consists of the following: a fiber chuck (Newport) to grip the fiber without stress damage, an xy miniature translation stage (Newport) to move the chuck vertically and side to side to capture a dispersed wavelength, and homemade aluminum brackets attached to opposite sides of each translation stage to grip the chuck and for attachment to the table. The brace directions are reversible to facilitate various launch stage geometries. This last point is important because the translation stages are the bulkiest part of the launch stages, and their placement can be staggered while keeping the fiber chucks in a relatively uniform plane of focus. This stage design is an important component in reducing the table area for the excitation system. The polished ends of the optical fibers are placed in front of the focused beam path or slightly away from the focus to minimize laser damage of each desired wavelength dispersed from the prisms (Figure A6).

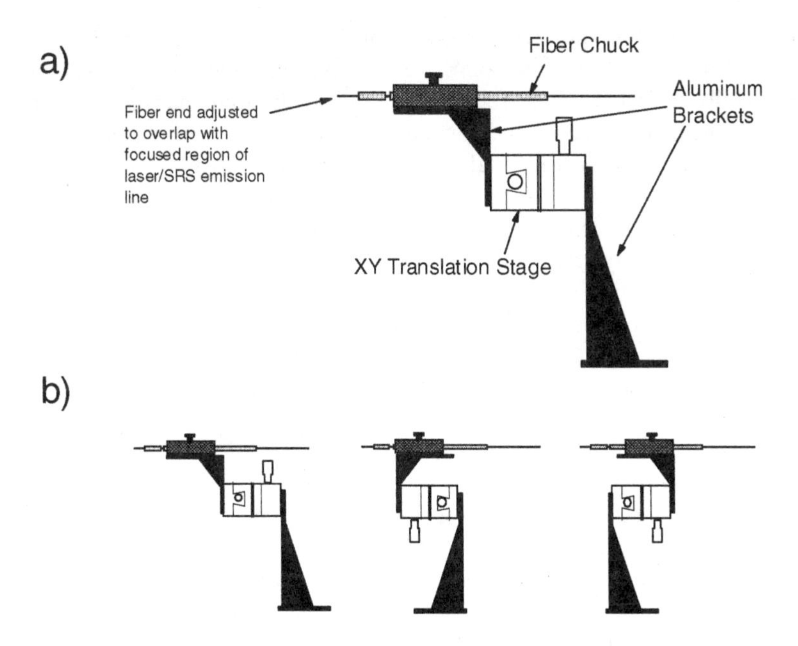

Figure A6. Optical fiber launch stage design (a) and various configurations available through manipulation of brackets (b).

A.4.2 Optical Fibers and Light Delivery

Optical fiber stock was purchased from Polymicro Technologies, Inc. The selected optical fiber is step index and multimode with a numerical aperture of 0.22. The inner core diameter is 400 μm; outer core diameter is 440 μm; and buffer diameter is 480 μm. The fiber is made of high-UV-transmitting fused silica, and the buffer layer is a polyimide coating for flexibility and resistance to damage.

With capability of up to 12 channels, each with an excitation and detection fiber, there are a total of 24 fibers brought to the LIF-CPT probe to monitor soil fluorescence. Cabling of these fibers is necessary to protect them from the rigors associated with remote monitoring. Four individual fibers are bundled into each cable sheath, resulting in a total of six cables to fit into the hollow space of the CPT rod. The cabling material used is 3.8-mm diameter "Red Standard Furcation" available from Northern Lights Cable, Inc. It has three layers to protect fibers, including an inner polyethylene jacket [inner diameter (ID) = 1.2 mm], a Kevlar yarn layer, and an outer polyurethane jacket. The four optical fibers are manually fed into the polyethylene jacket. Cables are bundled together using electrical tape and then coiled.

For the six cables connected to the probe, the diameter of the bundle is just under 0.5-in, which can be accommodated in a CPT rod with 1-in ID, bearing in mind that other electrical wiring for geophysical probes will also be threaded through the rod segments. On the end of the cables where exposed individual fibers break out into connector ends, smaller cable sheaths protect those individual fibers. These are the same construction as described previously but have an outer diameter (OD) of 3.0 mm ("Orange Standard Furcation" from Northern Lights Cable, Inc. Each four-fiber cable generally has four breakout cables at either end. The breakout portion is protected by a hard plastic covering, a polypropylene pipette tip with point removed, fixed by means of electrical tape and heat-shrink tubing.

There are essentially three different partitions of fiber cable in the LIF system. One partition connected to the laser excitation table collects excitation light, the second partition connected to the probe is actually threaded within the CPT rods, and the third partition is connected to the detection system. The partitions are joined on a patch panel using ST type connectors and adapters. The first and third partitions, instrumental optical fibers, are relatively short, not longer than the length of the CPT vehicle working area of 4 m, whereas the second partition of probe fibers is sufficiently long to access anticipated contaminant depths below the ground surface. Because of the way the CPT rod segments are stacked in the vehicle, there must be additional fiber length allowed to account for bending slack, about 0.5 m for every meter of rod segment. There is also the necessary length for the clearance between the ground surface and the hydraulic push system within the CPT vehicle and length for slack between the hydraulic push system and the patch panel. The total length for the probe cable is thus approximated as follows:

$$\text{Cable length} = [2 \times (\text{contaminant depth})] + 7\text{ m}$$

Optical fiber connectors are ST type and made by Augat. Connectors for the 400/440/480 multimode type fibers are specially overdrilled because standard connectors can accommodate a maximum 125-μm fiber diameter. These overdrilled connectors are obtained from a local supplier, The Fiber Optic Center, in New Bedford, Massachusetts. The coupling losses associated with these connectors are ~1 dB, and they drop to ~0.2 dB if index matching fluid is used at connector. Partitions of cables are connected by means of ST connectors attached via adapters in a connector panel. Standard procedures are followed in assembling connectors onto fiber cables. The connectors are bulkier than the cable bundles, so caution is used when threading probe cables through CPT rod segments. The lengths of the breakout cables are also staggered to offset the resultant connector positions (Figure A7).

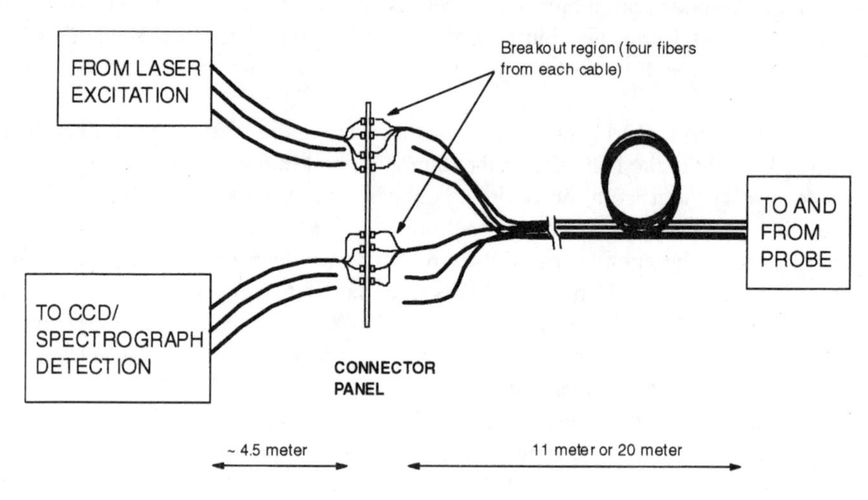

Figure A7. Layout of optical fiber cables used in LIF-EEM-CPT. Breakouts are only specifically presented for two of the connected cable pairs.

A feature affecting data acquisition is the interference associated with laser backreflection and backscatter. In soil samples where soil particles are reflecting, the amount of back-reflected light is substantial compared with actual fluorescence of contaminants. Although this back scatter may be addressed by data reduction procedures, the possibility of detector saturation may make it necessary to filter out backscattered light while passing fluorescence. This is achieved in the LIF-CPT tool by inserting a small cut-off filter in the fiber optic connector adapter between the connector ends. The appropriate type of filter depends on the channel of laser light being filtered. There are two disadvantages associated with this procedure. Total signal throughput is lowered due to the separation of the connector ends brought about by the width of the filter and the filter will often block some of the lower wavelength fluorescence response, depending on the PAH in question. The current probe was designed to minimize instrumental back-reflection from the probe window surfaces, so the need for filters will be reduced. However, should total signal be sufficiently high, their inclusion would still represent an instrumental advantage.

A.4.3 LIF Probe and Sampling Interface

The most significant changes in the LIF-EEM-CPT instrument described here from previous versions are associated with the probe design. There are two aspects that stand out.

1. The orientation of the excitation and detection optical fibers in each channel at the probe interface has been altered to minimize back-reflection from sapphire window surfaces.
2. The distance between channels has been dramatically reduced to lower the required sampling area to within the geophysical depth resolution standard of 2 cm.

The probe fits into a stainless steel CPT subassembly, or "sub," near the cone tip when it is pushed into the subsurface. It includes a transparent window that is in contact with subsurface soil, which provides both a means for rigidly holding the optical fibers in correct orientation against the inside window surface and protection for optical fibers or cables extending up the sub segment. The window in LIF-CPT technology is usually made of sapphire, which is sufficiently hard to resist damage and abrasion by soil particles under pressure from the rod. In other respects, such as attachment of the window to the sub and the design of the fiber holder, most previous designs only held a single pair of fibers. Even previous multiple-channel LIF-CPT probes had 10 such individual fiber pairs in its sub, each separated by $1^1/_2$ in along a vertical axis of the sub.

The progression of two-fiber probes, where one fiber acts as an excitation channel and the other fiber is the detection channel, is described in a recent paper by Lin et al.[5] Each stage represents an improvement in the collection efficiency of signal in the probe due to maximization of the overlap between the excitation and detection regions of the fibers on the outer sapphire window surface. Numerical aperture, window material, and window thickness dictate the path of light rays that affect overlap. The most recent Tufts two-fiber arrangement maintains the overlap of the excitation and detection surfaces but eliminates collection of back-reflected light arising between the sapphire window and the sampling area. Because reflection coefficients return a much higher proportion of excitation light than fluorescing molecules after absorption and quantum efficiency factors are considered, the elimination of such back-reflection removes a significant interference from the sample fluorescence. However, back-reflection from soil particles and Rayleigh scattering still occurs to the extent that backscattered light is still significant, but the degree of overall scattering is reduced.

In using multiple fiber pairs the excitation/detection surfaces for each channel must be separated from those for the other channels. The presence of overlap results in sample interface cross talk that corrupts the information provided in LIF-EEMs. The previous Tufts probe design prevented this problem by separating the channels a large distance, 1.5 in. This separation occurred vertically so that each channel sampled different subsurface depths, and data were depth corrected after sample acquisition. In the current probe design, the distance between each channel is closer to 2 mm to allow better depth resolution without correction. Through calculation and performance testing it is found that cross talk from channel proximity does not occur. Back-reflected and backscattered excitation light do remain strong after multiple internal reflections, but these cross talk features are narrow and can be subtracted from each channel fluorescence in LIF-EEMs.

To provide rigid angled orientation of the fibers and to allow the proximal mounting of 12 channels within the span of a single sapphire window, a "button" was designed. It is essentially a small solid cylinder of aluminum that is drilled with 24 small holes that are just larger than the optical fiber diameter (including buffer) of 0.480 mm. The button acts as a rigid guide for angular orientation as well as a platform for mounting multiple fibers. Holes are drilled by means of electric discharge machining (EDM). Fibers are fixed into the button by hard epoxy (Tra-Con BA-F113) used for optical fiber connectors. At the surface of the button in contact with the sapphire window small cups are machined about each hole pair to help capture epoxy at the surface. Protruding fibers and epoxy are then polished down flush with the original button surface.

The button is fixed to a clamp with a hollowed out area to accommodate bare fibers extending from the back of the button. The clamp comes in two halves that are screwed together, and the button is held by a flange in the clamp that fits around the button when the clamp is assembled. The faceplate with sapphire window is attached to the button/clamp assembly by screws. This three-component assembly is fixed to the sub walls by screws in the faceplate. In this way, the entire optical probe is rigidly fixed and to withstand the rigors of soil penetration. No springs are used, which should prevent disconnection between the window and the optical fiber ends. Therefore, the most crucial tolerance in this design is the distance between the inner sapphire window surface and the button surface, which relies on the button width tolerance, the window width tolerance, and the faceplate aperture tolerance. With these factors considered, the machine tolerance between the window and button surfaces is –0.000 to +0.002-in and is considered adequate.

Another issue with the probe assembly is the assurance of a leak free environment within the clamp/button/faceplate assembly. This is accomplished by a viton o-ring seal between the button and faceplate and by using sufficient epoxy when attaching the window to the faceplate to eliminate small channels or pores at that contact. A low-viscosity epoxy (Tra-Con F114) appears to provide sufficient wetting of surfaces to provide a leak-free window assembly.

The sub comes with two threaded attachments, one to attach to the CPT tip placed directly below and the other to be integrated with other CPT rods. The tip contains geophysical probes used during a LIF-CPT sounding. A standard CPT rod of 1.44-in OD or smaller is not able to accommodate fiber cables and gauge wires to probes at the tip, so the LIF probe is threaded into larger "BAT" pipe with 1.75-in OD.

The design necessitates connection of the LIF-CPT sub with the tip prior to attachment of the actual LIF-CPT probe due to lack of room for bulkier gauge wire ends associated with the tip probes once the LIF probe is in place. Therefore, the gauge wires are in place prior to attachment of the probe, and connector ends of the probe fibers are fed through the sub aperture and on through the rod segments required to reach a specified depth before screwing it into the sub (Figure A8).

A.4.4 Detection Instrumentation

To attain multiple channel fluorescence spectra, the detection system is comprised of the following components: an imaging spectrograph, optical fiber interface, and a CCD detector. The spectrograph and detector system were obtained as an integrated system from Instruments, SA (ISA). They are operated by the same Windows™ environment software built around the Grams 386 system, SpectraMax for Windows. The utilization of detection instrumentation operation in non-DOS format provides for multitasking and brings overall data acquisition closer to the model of comprehensive control through virtual instrumentation programming.

The *imaging spectrograph* is a 270M model manufactured by ISA and has a 270-mm focal length with nominal 3.1 nm/mm reciprocal linear dispersion with a 1200-g/mm grating. The spectrograph is larger and has better imaging capabilities than in previous LIF-CPT systems for two purposes. The first is that the CCD detector chip is unusually large, which requires that the spectrograph optics exhibit greater freedom from field curvature distortion. Second, a source of cross talk between channels arises from blurring of information along the vertical dimension of the CCD; better optical imaging reduces the likelihood of such blurring.

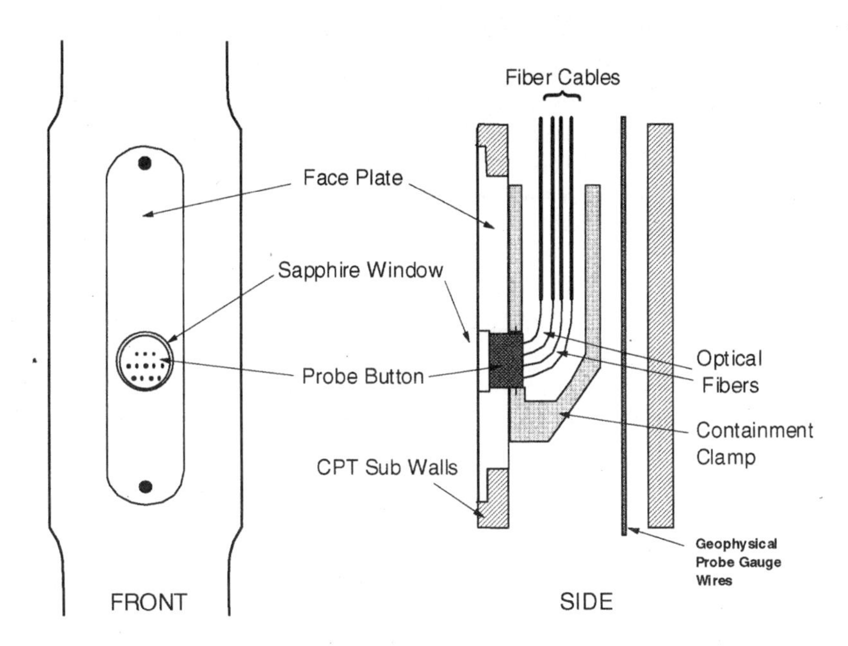

Figure A8. Probe assembly and CPT sub. Individual fiber pairs are seen on the front view of the probe button.

The spectrograph accommodates two gratings placed on opposite sides of a turret platform and can be conveniently switched by software commands. One of the selected gratings is a 600 g/mm grating blazed at 1000 nm for acquisition of Raman response from soils. For the first field operations, a 200 g/mm blazed at 726.6 nm was used for fluorescence detection. Since then, this grating was replaced with a 300-g/mm blazed at 250 nm. The first grating was intended to acquire fluorescence spectra in the second order, but significant overlap was found in the second order response with the first and third orders. The new grating provides first order response without the overlap problem.

The spectrograph has two input ports and two output ports. For the input ports, one is the entrance slit aperture, and the other is blocked; for the output ports, one accommodates the CCD detector, and the other is blocked. The extra port on each of the input and output of the spectrograph could be utilized for future applications. All facets of the spectrograph, including slit widths, center wavelength location, and grating selection, are controllable with the SpectraMax for Windows software.

The *CCD* was also acquired from ISA configured for spectral or imaging applications with the SpectraMax for Windows software. The chip has 2000 × 800 pixels, each 0.015 mm in size, supplied to ISA by Scientific Imaging Technologies (SITe), a principal manufacturer of such chips. The chip is thermoelectrically cooled and back thinned with UV enhanced antireflection coating to monitor fluorescence below 300 nm; the high wavelength cutoff is approximately 1000 nm. The chip's peak quantum efficiency is about 65%, whereas standard detectors without back thinning are at about 45%. It has a larger active area than other comparable CCD chips, allowing more resolved imaging with less likelihood of cross talk. The operating software provides sufficient binning to separate the 12 channels of information, where each channel has a variable superpixel width. The software also has a linearization feature that creates files with a calibrated wavelength x-axis.

In the *optical fiber interface*, 12 fibers are arranged vertically in alignment with the entrance slits of the spectrograph by means of an adapter plug designed at Tufts. The goal is to advance the fiber ends close to the plane of the slits and to provide alignment to optimize throughput for the fiber channels. Alternatively, an optical lens system might have been used to obtain a focused image of optical fiber ends on the entrance slit. The actual adapter design consists of a flange attached to the spectrograph and into which the plug is inserted. The flange also contains set screws to secure the adapter plug. The plug is comprised of two half pieces, each milled to snugly fit the optical fibers when the two halves are tightened together. To achieve appropriate spacing and separation of the fluorescence signals imaged by the spectrograph onto the CCD, the detection fibers are laid down on one of the halves separated by a "blank" fiber 200 μm in diameter and as long as the plug. The fiber arrangement is then covered with epoxy, and the other half of the plug tightened down. After the epoxy dries, the fiber ends are polished in the usual manner and inspected to ensure throughput for each channel.

A.5 SYSTEM PERFORMANCE

This section presents aspects of performance over the course of lab testing and field operations. Continuing instrumental development and system operation will influence measured values reported here.

A.5.1 Excitation System

Two wavelengths, 266 and 355 nm, are simultaneously delivered from the Nd:YAG laser source. Measurement of individual wavelengths after beam separation indicates that 41 to 48% of total measured power corresponds to 266 nm, and 52 to 59% corresponds to 355 nm, when both harmonic modules are optimally tuned.

Laser power was measured with a thermopile detector and radiometer manufactured and calibrated by LaserProbe Inc (RkT-10-PB for the probe, Rm-6600 for the radiometer). SRS pulse energies were measured with an RjP-637 energy probe, also used with Rm-6600. Note that measured source laser powers are average powers and can be converted to pulse energy by dividing the power by the repetition rate, 50 Hz.

When the harmonics are optimally tuned and flashlamps have not aged, total output of the laser is about 700 mW. For measurement of SRS outputs, each line after dispersion was directed into a 4-m optical fiber in the manner explained previously, and output was measured at the ends of each fiber. One such output measurement is presented in Table 2.

Table 2. Measured SRS Powers through Four Meter Optical Fiber

Wavelength (nm)	SRS Peak Assignment (*F, H, M*)	Measured Power
246.9	4, 0, 1	0.32 μJ
257.5	4, 1, −1	0.32
266.0	4, 0, 0	6.8
288.4	4, 0, −1	15.0
299.1	4, −1, 0	26.0
314.8	4, 0, −2	2.6
327.6	4, −1, −1	1.1
341.5	4, −2, 0	1.9
354.7	3, 0, 0	29.0
416.0	3, −1, 0	6.5

Such results are representative but subject to fluctuation. Because SRS lines are competing for available light flux, it is typical to see some lines go up in power while other lines go down. Furthermore, optical alignment may favor optimization of some wavelengths over others. Over the course of time, varying outputs that are experienced are due to the following considerations:

1. Flash lamp aging in the Nd:YAG laser is a long-term effect that occurs over months of operation. It is recommended to bring fresh flash lamps for field operations.
2. Solarization, fiber damage caused by light transmission, occurs relatively quickly in optical fibers but soon reaches an asymptote, whereupon further degradation occurs very slowly. It is recommended to allow solarization to occur prior to field operations to avoid output changes.
3. Alignment operations require upkeep on the order of hours or days, depending on temperature and mechanical stability of the operating environment. During field operations, alignment may be required between every successive LIF-CPT push. There are two types of alignment, including optimization of the harmonic tuning for 266 and 355 nm at the source laser and optimization of the fiber capture of SRS light. When alignment alterations occur, the change in output is sudden and may be dramatic, so it is impossible to correlate measurements that are separated in time by an alignment alteration. It is, therefore, necessary to reestablish system output levels after every alteration.
4. Mode instabilities are short term (an hour or less) and are represented by nonsystematic drift in the measured power. The relative variation may be as high as 50% of the total output in a channel; shorter-term pulse fluctuation noise is typically around 15%. It is noticeable when powers are measured through a fiber due to variation in the focused beam profile being captured by the fiber and to variation in the efficiency of the propagation of drifting modes within the fiber.

Long-term variation was also investigated. In this case, laser outputs were measured for each channel, except 354.7 and 416.0 nm, over the course of a day. At various times, the output from a 4-m optical fiber was measured for the various SRS excitation lines, in each case using six readings of 50 pulse averaged readout from the energy meter. Steady performance was obtained for each channel, with a variation over 8 hours of 20 to 40% for the least stable channels (314.8, 327.6, and 341.5 nm) and ~10% for the most stable channels (266.0, 288.4, and 299.1 nm) (Figure A9).

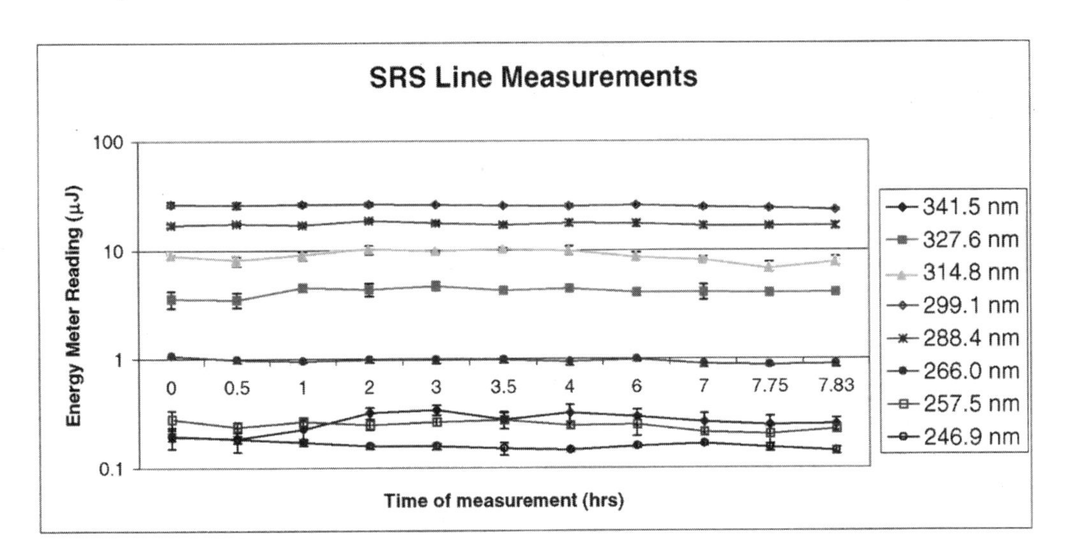

Figure A9. Long-term output power variation for various SRS excitation wavelengths.

The powers measured through optical fibers are expected to be lower than if they were measured directly on the excitation table. Although propagation through fibers exhibits some attenuation, the main source of lessened output arises from incomplete coupling of SRS light into the fiber. In cases of the strongest excitation wavelengths, 266, 355, and to a lesser extent, 288 and 299 nm, it is necessary to put the fiber in a defocused region of the laser path to avoid laser-induced damage. In other cases, focused excitation does not provide a particularly small or circular spot, so capture is not efficient. This occurs in the second stage of fibers at higher wavelengths (299 nm and above) because SRS is now passing through two lenses that can be placed correctly for one wavelength, and remaining wavelengths will be imperfectly propagated by the lenses.

In two field tests, it was found that laser source variation was greater in the field than in the controlled environment of the laboratory. The field environment is undoubtedly harsher for any instrument due to factors such as greater fluctuation of ambient temperatures, continuous operation over days, multiple power shutdowns, and constricted time schedule. Different laser systems will experience unique behaviors under such conditions.

A.5.2 LIF–CPT Probe

Measured excitation outputs, at four assembled probes, measured immediately prior to the first field test, are given in Table 3.

Table 3. Measured Powers of Excitation Lines from Assembled LIF-CPT Probes (All Values in μJ)

Wavelength (nm)	Probe A (11 m)	Probe B (18 m)	Probe C (11 m)	Probe D (18 m)
266.0	12.0	3.2	1.7	1.0
288.4	1.1	1.8	1.6	1.2
299.1	6.5	3.3	13.9	5.3
314.8	0.13	0.26	.42	0.12
341.5	.18	.66	0.44	0.49
354.7	9.0	7.4	5.6	3.6
416.0	2.7	6.9	4.3	1.8

Lengths of fibers listed for each probe are measured from the connector panel, so 4 m of additional fiber need to be considered for the Table 3 data. Data in Table 3 were collected during a calibration procedure where 246.9-, 257.7-, and 327.6-nm outputs were not rigorously measured due to their low energies of less than 0.1 μJ. Note that measured powers in Table 3 do not show consistency with one another, which is due to the aspects described in the previous section. These measurements occurred over a period of days with a number of alignment operations. However, general correlation is seen regarding lower output and longer cable length.

It is difficult to estimate the minimum energies required for obtaining signal fluorescence because of varying soil conditions, wavelength response, and efficiency for different PAH compounds, but the excitation powers expressed in Table 3 have been found to be sufficient for detecting ppm levels of any given PAH in soil. Probe D has been used in the field to successfully detect naphthalene and 2-methyl naphthalene at 7.8 ppm. In the laboratory, probe B was used to detect various levels of fluoranthene (16, 32.5, and 77 ppm, each with 7.5% moisture content). In loosely packed soil, 16 ppm was easily detected with 266.0, 288.4, 299.1, and 314.8 nm and to a lesser extent 341.5 nm. However, in packed soil, 32.5 ppm was detected with lower fluorescence at all wavelengths, to the extent that 314.8-nm fluorescence was barely seen and 341.5 nm fluorescence not seen at all. The CCD integration time was 0.3 sec, and the slits were set at 0.3 mm. In the field, integration time is currently established at 0.4 sec.

Whereas Table 3 provides results obtained at the LIF-CPT probes, these data generally reflect more on the excitation system characteristics. A crucial aspect of probe performance occurs in the correlation of throughput excitation energy and fluorescence response. Not only does this measurement include aspects of attenuation through excitation fibers, but it also includes detection fiber attenuation and overlap of the excitation and detection surfaces at the sapphire window/sample interface. Experiments were conducted for each probe that monitored the linear fluorescence detection response of Rhodamine B solution with different excitation energies for each excitation channel. This procedure is described in detail in the Operation/Calibration Manual included in

Appendix B. It was found that, for any given wavelength, the slope between measured excitation energy at the probe and measured rhodamine B fluorescence at the CCD/spectrograph differed greatly for the four probes (Figure A10). Furthermore, there was no clear correlation between the slopes and the detection fiber length. Because configuration at the detector interface was similar for the measurements, the remaining difference between the probes arises from different overlaps of excitation and detection surfaces. This variation is due to poor tolerances associated with the drilled holes in the probe button.

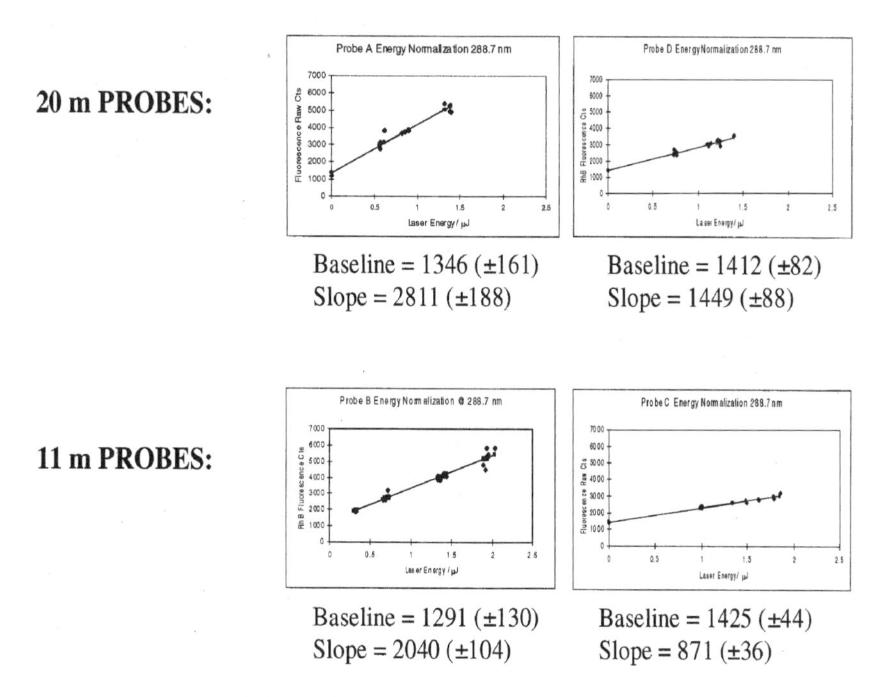

Figure A10. Response curves of rhodamine emission with 288-nm excitation energy for probes having two different optical fiber lengths.

Cross talk between excitation channel information at the probe/sample interface was also tested. Backscatter and back-reflected excitation light at certain wavelengths tends to be detected by multiple detection channels, and these features are easily subtracted from intrinsic fluorescence response in each channel. Numerous experiments reveal no fluorescence cross talk between channels.

A.5.3 Detector/Fiber Interface

The numerical aperture of the spectrograph system is 0.125, and the numerical aperture of the fiber output is 0.22. It is clear that some light being emitted from the detection fibers will overfill the spectrograph and be lost. Second, the entrance slits, typically at 0.3 mm, will block light from fibers whose effective diameters are 0.4 mm. The fiber adapter plug described in Section A.4.2 was designed to bring the fibers quite close to the slits, thereby minimizing the light lost at that point. Losses associated with this fiber/spectrograph interface are considered acceptable given the advantages of the simplified adapter plug configuration.

A.5.4 CCD Detector Performance

Performance of CCD detection has been satisfactory. Two particular aspects bear mention, including the binning operations with channel separation and the detector background response.

Determination of binning was based on a CCD image of the vertical array of fibers each transmitting the Hg lamp line spectrum. Based on this image, which showed the fiber outputs were not fully resolved on the CCD, vertically illuminated pixel ranges were identified and binned accordingly for each fiber channel using the detector software. Each binned portion was separated from its upper and lower neighbors by approximately 50 pixels. These pixels contained overlapping information from the adjacent channels and thus were excluded from the CCD readout. In fact, each channel contained only eight pixels of actual readout in the vertical direction, although 800 total pixels are available with the CCD detector. The resulting multichannel spectra were completely free of detector cross talk arising from image blurring.

During laboratory diagnostics and experiments, the CCD background counts arising from spontaneous generation of charges were found to be fairly uniform across the wavelength axis of the CCD for each channel. For 0.4-sec integration time, the count level was approximately 1200. Throughout the first field operation, the background counts were not uniform and showed structure. This resulted in difficulty in developing automated routines for subtracting background counts from actual fluorescence signal counts. When fluorescence response was low, baseline structure was found to be comparable in magnitude to fluorescence structure. In the second field test and during laboratory operations, the baseline did not show structure.

REFERENCES

1. J. Lin, S.J. Hart, T.A. Taylor, J.E. Kenny. 1994. "Laser fluorescence EEM probe for cone penetrometer pollution analysis," *SPIE/A&WMA Symposium on Optical Sensing for Environmental and Process Monitoring*, Vol. 2367, p. 70–79.
2. S.J. Hart, Y-M. Chen, J.E. Kenny, B.K. Lien, T.W. Best. "Field demonstration of a multichannel fiber optic laser induced fluorescence system in a cone penetrometer vehicle," *Field Anal. Chem. Technol.*, 1(6), 343-355.
3. G.B. Jarvis, S. Mathew, J.E. Kenny. 1994. "Evaluation of Nd:YAG-pumped Raman shifter as a broad-spectrum light source," *Applied Optics,* 33(21), 4938.
4. S.W. Allison, G.T. Gillies, D.W. Magnuson, T.S. Pagano. 1985. "Pulsed laser damage to optical fibers," *Applied Optics,* 24(19), 3140.
5. J. Lin, S.J. Hart, J.E. Kenny. 1997. "Improved two-fiber probe for in situ spectroscopic measurements," *Anal. Chem.,* 68(18), 3098.

APPENDIX B

Operation and Calibration Manual for the Laser Excitation–Emission Matrix Cone Penetrometer Tool

Jonathan E. Kenny, Jane W. Pepper, and Andrew O. Wright

B.1 INTRODUCTION

This manual describes the assembly, calibration, and operation procedures for a laser-induced fluorescence (LIF) excitation–emission matrix (EEM) tool used with cone penetrometer technology (CPT) in detection of subsurface contaminants, such as polyaromatic hydrocarbons (PAHs). This manual is complementary to the Design Manual (appended as Appendix A), which provides the ideas behind the design of a LIF-EEM-CPT tool that was assembled and tested by the scientists at Tufts University (Medford, MA). The theories of lasers and stimulated Raman scattering (SRS), light throughput in optical fibers, and optical detection systems are discussed in detail in the Design Manual, which was submitted to the U.S. Department of Defense (DOD) Advanced Applied Technology Demonstration Facility (AATDF) at the Energy and Environmental Systems Institutes of Rice University.

In recent years LIF coupled with CPT technology has been used widely as a highly sensitive and selective screening tool for underground contaminants.[1-5] Compared to conventional laboratory analytical techniques such as gas chromatography/mass spectroscopy (GC/MS) and high-performance liquid chromatography (HPLC), LIF-CPT has proven to be rapid, cost effective, provides real-time information without performing sample extraction and eliminates the turnaround periods required for laboratory analysis.

Conventional LIF techniques often focus on single wavelength excitation or the collection of total emission signals without distinction of the fingerprints of different fluorescing species. Greater details of individual compounds in a mixture can be provided by means of EEMs, a multichannel excitation–emission presentation based on the intrinsic wavelength dependence for both excitation and emission of a compound. Multiple excitation beams can be generated with a Nd:YAG laser-pumped Raman shifter filled with a gas mixture of hydrogen and methane, using laser stimulated Raman scattering.[6] Emissions from all the excitation wavelengths can be detected simultaneously with a charge-coupled device (CCD) detector. With this fluorescence EEM methodology, particular compounds or classes of compounds can be distinguished.[7]

Two generations of LIF-EEM-CPT tools have been designed, assembled, and tested at Tufts University. They were both equipped with 10 excitation and emission channels. The first instrument[8] was assembled into a cone rod with 10 sapphire windows, each window corresponding to one excitation–emission fiber pair, located 1.5-in apart from each other. It was assembled into a truck similar to that used in the Tri-Service Characterization and Analysis Penetrometer System (SCAPS). The tool was used successfully in a project funded by a grant under U.S. Environmental Protection Agency (EPA) for ground contamination assessment at Hanscom Air Force Base (AFB), Massachusetts, and at U.S. Coast Guard Station in Elizabeth City, North Carolina.

To improve the capabilities of the EPA LIF tool, a second generation LIF system was constructed with AATDF funding with the following goals:

1. Extending the overall range of excitation wavelengths available for better separation of PAHs
2. Extending the overall sensitivity of detection through higher excitation energies, better light collection, and reduction of the instrument artifacts
3. Reducing the sampling area at the probe/soil interface for better correlation of the signals collected within the 10 excitation–emission channels and for better depth resolution of each CPT operation
4. Reducing the space required for instrument deployment in a CPT vehicle to make it more field accessible
5. Allowing real time and automatic data collection and streamlining the data analysis process

The second-generation LIF-EEM tool has been assembled and installed in a CPT truck supplied by Fugro Geosciences, Inc. This probe allows real-time collection of laser-induced fluorescence EEMs in soils and in solutions. It provides spectral information to qualitatively and semiquantitatively assess subsurface contaminants such as PAHs and benzene, toluene, ethylbenzene, and xylenes (BTEX) at concentrations from neat liquids to parts per million or less at depths up to 50 ft. The probe was tested, calibrated, and then deployed to characterize subsurface soil contamination at test sites located at Hanscom AFB, near Boston, and Otis Air National Guard Base (OANGB) in Falmouth, MA.

B.2 ASSEMBLY OF LIF/CPT SYSTEM

B.2.1 Fiber Probe Assembly

B.2.1.1 Material

Fused silica step-index optical fibers used for this LIF-CPT application are 400 μm in core diameter, with additional 20-μm thick-cladding and 20-μm-thick polyimide buffer. They are multimode, with a numerical aperture (NA) of 0.22 (see Design Manual, Appendix A). The UV transmissive 400/440/480 size fibers are purchased from Polymicro® Technologies. Depending on the depth of the contaminants to be detected, fiber probes of different lengths are assembled. Because of the way the CPT rod segments are stacked in the vehicle, additional fiber lengths are necessary to cover the bending slack. More fiber length is also required to account for the distances from the ground surface to the hydraulic push system within the CPT vehicle and from the hydraulic system to the patch panel. The total necessary fiber length is thus more than double the achievable depth for detecting subsurface contaminants. The length can be calculated using the following equation:

$$\text{Fiber length} = [2 \times (\text{desired push depth})] + 7\ \text{m}$$

In this way we constructed two probes with total lengths of 19 and 11 m to access depths of 6 m (20 ft) and 2 m (7 ft), respectively. Taking into consideration the possibilities that broken fiber probes might occur during a field operation, backup probes with the same lengths were assembled.

Standard bifurcation cables (Northern Light Cable, Inc.) used to protect the fibers are measured at slightly shorter lengths than the optical fibers, which leaves about 1 ft of bare fibers at each end.

These red cables have a Kevlar jacket of outer diameter (OD) ~1/7 in, and a rigid buffer of ~1/13 in inner diameter (ID). Strong string throughout the length of the cable can be used to pull the fiber through the cable. Four fibers with size of 480 μm OD can be placed into the "red cable." If more fibers, up to six, are placed in the same cable, the stress results in broken fibers when the cable is bent. If fewer than four fibers are used per cable, then more cables and more space are needed.

An orange cable from Northern Lights Cable, Inc. with a structure similar to the red cable but a smaller diameter (jacket OD ~1/9 in buffer ID ~1/25 in) can also be used to protect the fibers. A single fiber is usually placed within the orange cable.

A patch panel that can host up to 24 connectors can be purchased commercially. The panel serves to connect all fibers from the laser breadboard to the excitation fibers of the LIF probe and from the emission fibers of the probe to the detector.

ST™ multimode connectors with metal ferrules are purchased from Augat®. The connectors are specially drilled to fit the size of the optical fiber that is used in this application, with cladding OD of 440 μm. All items required for terminating the Augat® ST™ connectors can be obtained by purchasing the Augat ST Connector Tool Kit #68000-010-01.

To polish the fibers, lapping films (AngstromLap™) are used with aluminum oxide particle sizes of 12, 3, 1, and 0.3 μm. The 12-μm film is for coarse polishing, and the 0.3 μm is for extra fine polishing. The polishing tool and the glass plate onto which the lapping films will be adhered are cleaned prior to use to prevent fiber breakage during polishing. Drops of water are added to the lapping films while polishing.

Three types of epoxy were purchased from TRA-CON®, Tra-Bond BA-F113SC (black), BA-2170 (yellow), and BA-F114C (clear). They all cure at room temperature after 24 hours. The black epoxy also cures after 10 minutes in a 100°C oven. They are used for fiber-connector assembly, fiber probe assembly, filling the probe sub, and for gluing the sapphire window onto the face plate.

B.2.1.2 Assembly of Fiber Probe

Making the LIF Probe: Feed/Pull the Fibers through the Cable — Fibers and cables are laid out in a straight line. Four fibers are placed as a bundle to be pulled through one red cable. Because of the stress within the bundle, it is difficult to feed them through the cable that is 20 m long or longer. To pull fibers through the cable, four fibers are bundled to form a square cross-section, and a string is glued inside the first 2 in of the fiber bundle with Duco Cement®. The cement dries in about 5 minutes at room temperature and retains considerable strength. When the glue is dried, the string can be pulled through the cable to advance the fiber bundle. When finished, the string and the glued section of fibers are cut off and discarded.

One End of the Fibers Glued to a Button — Six cables with four fibers each (totally 24 fibers) are used in making 1 LIF probe, with 10 fiber pairs as 10 excitation–emission channels and the other 2 pairs as backup. The six cables are stretched out and taped into a bundle with electric tape. Coil up the bundle. At one end of the bundle, fibers are placed within the probe button (see "Probe Button" in Figure B1), which has 24 angled holes for the excitation–emission fiber pairs. With all the fiber ends reaching out of the button in an angle, apply black epoxy to the front of the button to glue the fiber pairs in place. Care is taken to have the enclosures at the button surface capture the epoxy and not to have too much epoxy.

Dry the epoxy over night and then polish the button. To make sure the surface of the button remains flat after polishing, the procedure follows a "figure 8" pattern movement. To speed up the polishing process, at the beginning, a piece of coarse sandpaper can be used to polish down the extra fiber and epoxy. This is done gently to avoid breaking/cracking of the fibers. When the epoxy is about 0.5 mm thick, switch to polishing with the lapping films of (in order) 12, 3, 1, and 0.3 μm. At the other end of the bundle, fibers are placed into ST connectors.

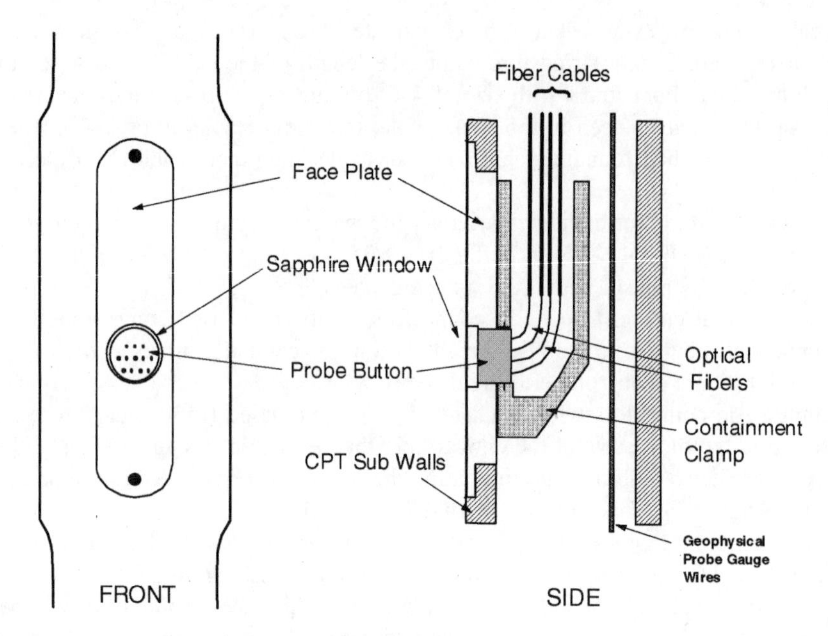

Figure B1. Probe assembly and CPT sub. Individual fiber pairs are seen on the front view of the probe button (Figure 9 in Design Manual).

Place Fibers into Connectors — The fibers on the other end of the red cable were left about 7 to 15 in long, with varied lengths for each fiber. Fiber breakouts are used to separate each of the fibers in a red cable and place them into individual orange cables. To do so, for each red cable, cut a rigid pipette tip to fit its smaller end to the red cable. Place the orange cables (with appropriate length for each fiber) through the fibers and fit all the orange cables into the bigger end of the pipette tip. Use heat-shrink tubing to tighten both ends of the pipette tip on the red cable and on the orange cables, respectively (Figure B2).

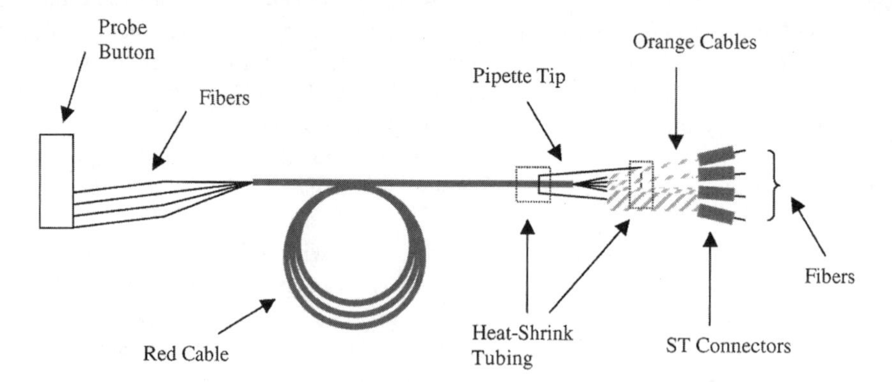

Figure B2. Schematic of fiber assembly of both ends: to the probe button and to the ST Connectors.

The fibers on the other end of the orange cables are left with about 1 in of bare fiber to be fitted into Augat® ST multi-mode connectors.[9] Fit fibers to connectors as follows:

1. Plug in the Augat multicure oven and set its temperature at 110°C.
2. Thoroughly mix the black epoxy in a pack and fill a 3cc syringe with the mixed epoxy.
3. Slide the ST connector's strain relief boot and crimp sleeve onto the fiber optic cable.

Strip the cable as shown in Figure B3.

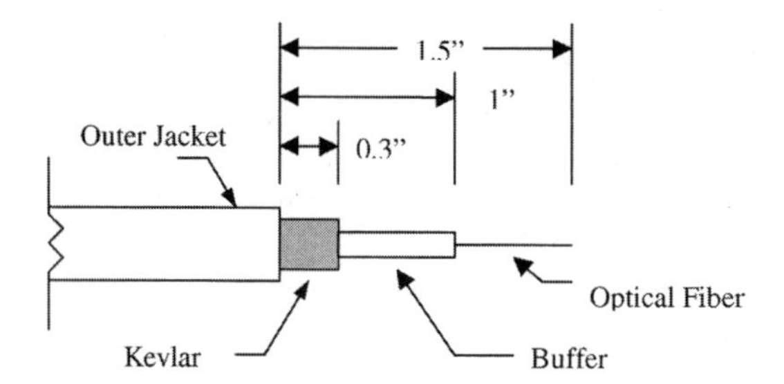

Figure B3. Schematic of an optical fiber in a fiber optic cable.

4. Burn the buffer of the fiber tip off (~$^1/_2$ in) with a cigarette lighter and clean the exposed fiber a piece of Kimwipe® soaked with alcohol.
5. Fill the connector with the black epoxy in the syringe, with a bead of epoxy forming on the tip of the connector (Figure B4).

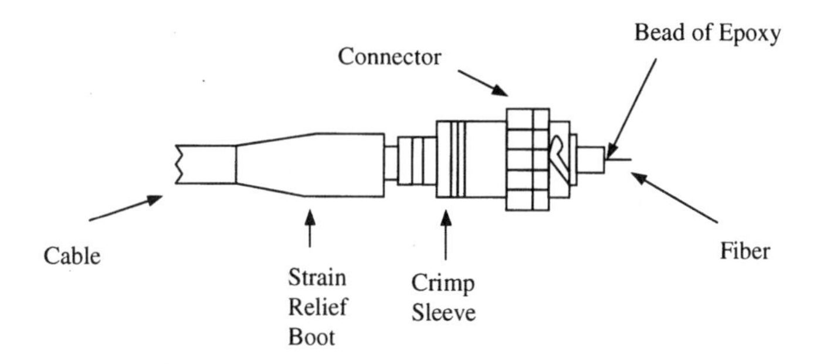

Figure B4. Placing a fiber in a connector and fitting the connector to a cable.

6. Insert the fiber into the connector, gently sliding the fiber in and out several times without completely removing the fiber from the connector for wicking of epoxy. Secure the connector under the cable between the fiber and the Kevlar.
7. Slide the crimp sleeve over the kelvar and connector backpost and crimp the sleeve using a crimp tool. Slide the strain relief boot over the crimp sleeve and connector.
8. Insert the connector into the ST heat sink and place it into the multicure oven for 10 minutes.
9. Gently cut off the excess fiber with a scribing tool. Polish the connector sequentially on 12-, 3-, 1-, 0.3-µm lapping films, using "figure 8" patterns.
10. Inspect the polished fiber under a microscope. Check the fiber by holding one end of the cable assembly to a light source and observing whether the light is visible at the other end.

Make certain that the lengths of the orange cables (among the four fibers within one red cable and among the six red cables) differ some so that the connectors are staggered, about 1/4 in apart from one another. Consequently, the whole diameter of the connector/cable bundle is smaller than the CPT rod ID (~7/8 in) and they can slide easily in and out of the rod segments.

Fit Button into the Sub — Leave ~1.5 in of bare fiber from the red cable end to the button end (within the length of the sub). Use Teflon tape to wrap around the bare fiber. Fit the button into the slot of the sub (in half). Bend fibers carefully along the direction of the angle (see Figure 1). Take care that fibers are not broken due to stress. Put on the other half piece of the sub and screw these two together. Make sure that the red cables are inside the sub. Tie the red cables to the handle on one of the sub piece with cable tie. Fully fill the whole sub with yellow epoxy (~10 packs). Set aside for 24 hours to cure.

Assembly of Face Plate — Place an o-ring on the button at the specific slot. Gently push the button into the face-plate (to avoid edging out the o-ring) until it is about $^1/_4$ in away from the rim of the face plate. Use a syringe to add a very little of the clear epoxy along the rim (housing of the sapphire window). Place an alcohol-cleaned sapphire window onto the rim and push down to make sure the sapphire window is fully within the face plate and that the surface of the face plate with the window is flat (see Figure 1 Side view). Add a small amount of epoxy to fill the gaps between the window and the face plate aperture. Wipe clean the epoxy. Set the whole piece aside for 24 hours to cure. When the window is stabilized, gently push the button all the way into the face-plate and screw tightly (see Design Manual Figure 9).

At this point, a fiber optic probe with sapphire window is fully assembled. Backup probes are assembled in the same way.

Other System Cabling — Addition cables are assembled to direct light from the laser breadboard to the patch panel and from the patch panel to the spectrograph.

Cable from Laser Table to Patch Panel — Measure 10-ft long fibers (12 in total) and fit four of them into one 8 foot long red cable. Fibers at one end of the red cable are fit into ~10-in long orange cables and then fitted into ST connectors. The procedures for fiber-connector assembly and polishing are shown in Section 2.1.2. The red and orange cables are connected with heat shrink tubing. Strip off the buffer of the fibers at the other end of the red cable (~1/3 in). Wipe clean with an alcohol pad. Place the fibers into each individual fiber chuck at the launch stage. Slide fibers in and out to have the fiber tips at the center spot of the excitation beam (for 266-, 299-, and 355-nm launching fibers, place them ~1/3 in back away from the focus spot to avoid laser-induced damage of the fiber). Turn screw on the chuck gently to tighten the fiber. Adjust the XY turret to position the fiber horizontally and vertically centered.

Cable from Patch Panel to the Spectrograph — Measure 12 fibers about 10 ft long and fit four of them into each of three 8 ft-long red cables. One end of the fibers is connected to the orange cable (~10 in) and then the ST connector as shown previously. The fibers on the other are to be glued onto a fiber/spectrograph adapter, which is specially designed and machined for the spectrograph 270M entrance.

The procedure of assembly is as follows:

1. Place the bare fiber (~2 in long each) on the center of the adapter. Place a 200-μm fiber as a spacer in between each optical fiber. Tape them in place with a piece of tape. Even the ends of the fiber tips and place them about 1/8 in beyond the adapter.
2. Screw down the other half of the adapter gently and firmly.
3. Use syringe to place several drops of epoxy on top of the fiber tip until a large bead is formed around all the fibers.
4. Set aside to cure for 24 hours.
5. Place the two angled metal pieces on the cylinder and screw down. Make sure the ends of the red cables are within the angled pieces to have the fibers protected.
6. Screw the cylinder into its outer housing, which is in turn screwed onto the spectrograph entrance.

When all the cables and the probe are assembled, place the fiber connectors onto the patch panel. There are 24 connector ports on the panel. The connectors of the fibers from the laser table are to be connected to one half of one side of the patch panel. The other half of the connector ports at the same side are to be connected to the fibers going to the spectrograph. The 24 connectors from the probe are to be connected on the other side of the panel. The 12 excitation fibers are connected to the fibers from the laser table, and the 12 emission fibers are connected to the fibers going to the spectrograph. Cutoff filters, with thickness no greater than 1 mm, for the necessary

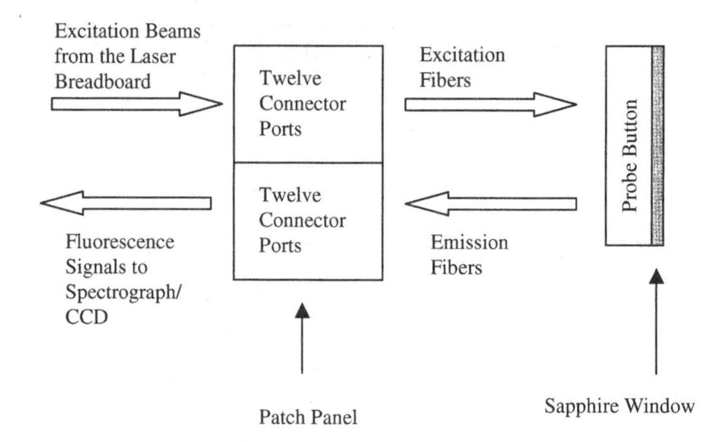

Figure B5. Diagram of setup at the patch panel.

channels to cut down laser backscattered excitation are placed between two connectors at the patch panel at the emission channel. Figure B5 shows the assembly scheme at the patch panel. Note that two excitation/emission fiber pairs at the probe are used as backup.

B.3 ASSEMBLY OF LASER BREADBOARD

The detailed design and assembly of the components of the laser breadboard are shown in the Design Manual. A Brilliant™ Nd:YAG laser was purchased from Quantel®. Its repetition rate is 50 Hz, with 266- and 355-nm output to pump a Raman shifter through the process of stimulated Raman scattering (SRS).[6] The resulting wavelengths are dispersed with a prism system. Ten of the wavelengths with the highest intensities are launched into the fibers. A schematic of the layout of the laser breadboard is shown in Figure B6. The procedures of the Raman shifter assembly and optics alignment on the breadboard are described herein. The detailed procedures of flashlamp and harmonic generation module installations and of laser alignment verification are described in the literature.[10] A summary of these procedures is presented as follows.

B.3.1 Installation of Flashlamp[10]

1. While the electronics and the circulating system are completely shut down, unplug the main power cord and push the capacitor discharge switch.
2. Remove the screws on the protective housing. Slide the protective housing toward the rear side of the laser optical head. Remove watertight nuts and screws on the pump cavity cover within the housing.
3. Identify the anode (+) and cathode (–) positions of the flashlamp to be mounted, and gently push its end wires through the watertight pieces with respect to the polarity indicted on the pump cavity cover.
4. Gently place the flashlamp in its bed in the laser optical head.
5. Tighten the flashlamp securing screws and the watertight nuts. Put back in place the pump cavity cover and the protective housing.
6. Pull the capacitor discharge switch and plug the main power cord.

B.3.2 Installation of Harmonic Generation (HG) Modules[10]

1. To assemble the second HG module with the oscillator, first remove the screws of the second HG module protective housing and pull it off.

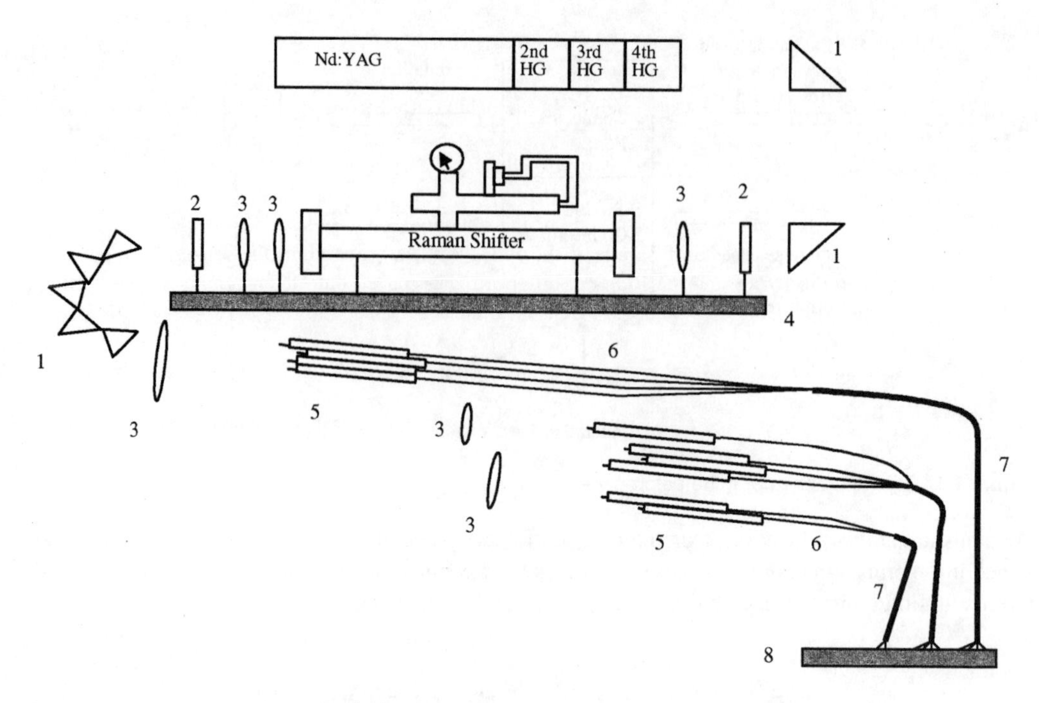

Figure B6. Schematic of laser breadboard. (1) Prism(s), (2) iris, (3) lens, (4) optical rail, (5) fiber chuck, (6) optical fiber, (7) cable, and (8) patch panel.

2. Hold the second HG module and adjust the position guide in front of the corresponding adapter of the oscillator. Assemble them together.
3. Tighten the screws on the module. Screw the supporting leg until it is in contact with the table, and lock it with the nut.
4. Put back the second HG module protective housing and tighten its screws.
5. Install the third and fourth HG module in the same manner (steps a to d immediately preceeding).

B.3.3 Laser Alignment Verification[10]

1. Start up the laser and let simmer current stabilize for 10 minutes (its voltage should be adjusted to 8.5 V for proper operation of the laser). Use the *Single Pulse* button of the Remote Control Box to operate the laser.
2. Place a piece of burn paper at approximately 1ft from the output port and fire a single pulse on the white side of the burn paper.
3. Compare the burn patterns to the one that was delivered with the laser. If no significant difference is observed, the laser is aligned.
4. The phase-matching adjustment knobs at the second and third harmonic modules and the thermal adjustment knob at the fourth harmonic module can be used to optimize the HG modules. With optimization and alignment, the output power for a new flashlamp is measured to be 700 mW with a power meter.

B.3.4 Assembly of Raman Shifter

1. Assemble all components of the Raman shifter (see Design Manual Figure 1a).
2. Purge the Raman shifter with methane gas for about 20 minutes.
3. Fill the Raman shifter with methane gas to the gauge pressure of 75 psi. Continue on filling the Raman shifter with hydrogen gas to final reading of 150 psi on the gauge.
4. Check for gas leakage and tighten all nuts as required.

B.3.5 Optics Alignment on the Optical Breadboard

1. All optics are adjusted to the same height as the beam output from the laser (see Design Manual Chapter III).
2. Adjust positions of the two prisms in front of the laser output to direct the 266 nm and 355 nm beams from the fourth HG to the optical rail.
3. With only the irises installed at both ends of the optical rail, the beams should pass through the center of both the irises, traveling parallel to the optical rail, before reaching the five-prism dispersing system. When the lenses and the Raman Shifter are later placed onto their mounts on the optical rail, their positions should be adjusted to ensure the laser beams still pass through both the irises and the collimated beams are directed to the five prisms.
4. The five prisms of the dispersing system are assembled at certain angles adjacent to each other using jigs. The positions of the prisms are stabilized on the mount firmly with double-sided tape. After dispersing, the beams from the Raman shifter are refracted approximately 180° and travel toward the center of the breadboard.
5. A focusing lens is used to launch the four lower wavelength beams (247, 258, 266, and 289 nm) to the fibers at the first launch stage (Figure 6).
6. Two other lenses with shorter focal lengths are used to focus the higher-wavelength beams (299, 315, 327, 341, 355, and 416 nm) to the fibers at the second launch stage.
7. A white card can be placed at the fiber ends to test the brightness (intensity) of the beams and the position of the fiber. The XY translation stages (see Design Manual) that hold the fibers are adjusted to locate each fiber at the center of the excitation beam.

When all components are aligned and optimized, the launch energies measured at the end of the excitation fiber of each wavelength are as indicated in Table 1 (see also Design Manual Table 2, Appendix A).

Table 1. Measured SRS Energy Outputs Through a Four-Meter Optical Fiber

λ (nm)	246.9	257.5	266	288.4	299.1	314.8	327.6	341.5	354.7	416.0
E (μJ)	0.32	0.32	6.80	15.00	26.00	2.60	1.10	1.90	29.00	6.50

After all components are assembled and aligned, only minute optimization and alignment are necessary on a day-to-day operation basis.

B.4 CALIBRATION OF LIF/CPT TOOL

B.4.1 Detection System Alignment and Calibration

B.4.1.1 Alignment and Focusing of the Detection System

Place the CCD [SpectrumOne™, Instrument SA, Inc., (ISA)] and the fiber/spectrograph adapter (with detection fibers aligned vertically) in position. The position of the head should be adjusted so that the CCD sensor lies at the instrument focal plane and that the images of the fibers are aligned with the pixel columns.

A Hg/Ar lamp (Oriel®) is used to check the alignment of the CCD and the fiber/spectrograph adapter, which dictates the positions of the fibers in front of the entrance slit.

This is done following these steps:

1. Place the lamp in front of the sapphire window (another position that can be used is in front of one of the connectors that connect fibers going into the spectrograph). Turn on the Hg lamp and let it warm up 10 minutes.

2. Use real time display (RTD) of the software (SpectraMax for Windows™, ISA) to take Hg spectrum. To have better resolution, smaller slit width (e.g., less than 0.050 mm) is necessary. Adjust the integration time to ensure that the detector is not saturated (i.e., highest intensity is far from the upper limit of 65,000 counts).
3. Move the spectrometer to a wavelength that ensures the highest intensity is at the center of the CCD array.
4. Move CCD gently in and out and rotate the CCD to obtain the sharpest peak possible for the centered peak. In between, also adjust the fiber/spectrograph adapter to make certain the fibers are aligned vertically at the entrance slit. Adjust these two elements and optimize both simultaneously until a sharp (less than seven pixels wide for half peak width) and symmetric peak is obtained. The following figures (Figure B7) show the cases of good and bad focusing and alignment.[11]
5. Check the other peaks that show intensity well above baseline. They should show the same focus and alignment. The focusing of those peaks detected at the edge of the CCD is less perfect compared to the ones located at the center of the CCD. However, the difference should be less than 5%.

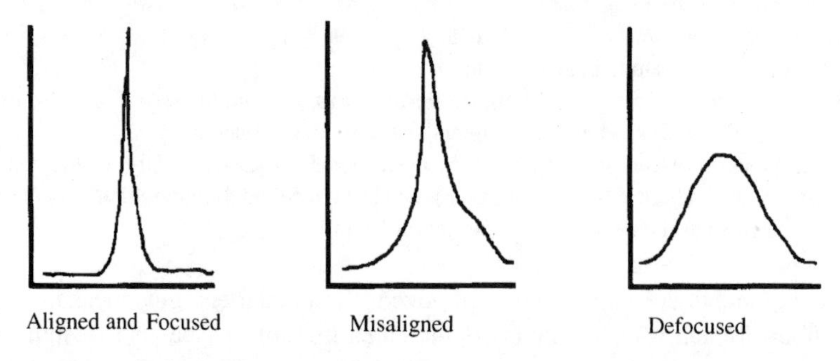

Figure B7. Examples of peaks during alignment and focusing. When aligned and focused, the peak is sharp and symmetric; when misaligned, the peak is asymmetric; when defocused, the peak is broad.

B.4.1.2 Imaging

Use the image measurement routine in the software to take the full image of the fibers. When the system is aligned and focused, the images of the fibers (12 rows of circles corresponding to the 12 detection channels and each circle corresponding to 1 Hg emission wavelength) are also aligned vertically and horizontally. When the CCD is not saturated (i.e., no "blooming" occurs) each image (i.e., each circle) should be fully separated to avoid cross talk at the detector.

The imaging is also useful to determine binning strategies. Binning involves combining photogenerated charge from adjacent pixels on the detector prior to readout. The binning options allowed by the software can improve signal-to-noise ratio, dynamic range, and readout speed. The trade-off is decreased spatial resolution.[12] Binning can be done by binning column pixels of each row (X-binning) or by binning row pixels of each column (Y-binning).

For Y-binning, take the images of the fibers and record the positions (i.e., the vertical pixel numbers) of each image. Choose the center pixels of a fiber image and group them into a superpixel for the detection of signals coming from that particular fiber (which represents a detection channel). Do the same binning for other channels. Signal that spills over to rows outside the range of the vertically binned strips are not processed. Again, the binning regions should be fully separated to avoid cross talk. With this Y-binning process, the detection area on the CCD can be divided into 12 distinctive strips, 1 for each emission fiber (10 channels are being used and 2 are backups). Save the CCD area file so that it can be retrieved later. For X-binning, all column pixels are used, and are binned into groups of 8.

B.4.1.3 Linearizing a CCD[13]

Each pixel on the CCD chip is sized 15 μm. Given the correct wavelengths of several peaks on a spectrum and the position of these peaks (in pixel number), the wavelength range that each pixel represents (nm/pixel) can be calculated. In turn, all the wavelengths of other peaks on the spectrum can be determined. This can be done manually or by software. A software (SpectraMax for Windows®) from ISA (Edison, NJ) is used for all data acquisition and wavelength calibration (i.e., linearization) of the CCD. In SPTWMAX/ISA_INI/ directory, view the file MONO1.INI, which controls the parameters for the linearization. Locate and record the value for the BASE_GRATING, OUTPUT_FOCAL_LENGTH, and FOCAL_PLANE_ANGLE. The calculated values of OUTPUT_FOCAL_LENGTH and FOCAL_PLANE_ANGLE by the linearization program are later inserted.

The detailed wavelength calibration procedure is shown in Reference 13, and a summary is described as follows:

1. With Hg lamp, choose the peak with the highest intensity and position it at the center of the CCD array (center pixel is 1000 when using a 2000 × 800 array). This is done by typing its wavelength to the box for "central wavelength." If the peak is not centered, adjust the spectrometer position until the peak is at the center.
2. Then set the calibration in the Instrument Control Center/Setup Application/Grating dialog box and choose the grating that is to be calibrated. Type in the wavelength of the center peak; click Calibrate.
3. Execute program Do Program/Linear.AB/One Scan-Multi Peaks. Then take the spectrum of Hg again.
4. Choose the five peaks that show the highest intensity and that span across the CCD detection region (to ensure the wavelengths in the whole CCD array are calibrated). Provide the correct wavelengths of these peaks to the software, which calculates the values of OUTPUT_FOCAL_LENGTH and FOCAL_PLANE_ANGLE.
5. Record these values and edit the MONO1.INI file accordingly. Save the file and restart the software before data acquisition.
6. Verify the linearization and repeat as necessary. When this step is done, the program will give the correct wavelengths of the spectra taken afterward.

To have the CCD show wavelengths when taking spectra, the "linearization" box in "CCD area" should be checked during data acquisition.

B.4.2 Calibration of the LIF Probe

B.4.2.1 Determination of Instrument Artifacts

For the multichannel LIF/EEM measurements, 10 different excitation wavelengths are used to monitor the fingerprints of fluorescing species. Fluorescence intensity, M, can be expressed as a function of excitation and emission wavelengths, λ_x and λ_m, respectively. The matrix elements, $M(\lambda_x, \lambda_m)$, of an EEM for a single fluorescing compound are given by Equation 1,

$$M(\lambda_x, \lambda_m) = kI(\lambda_x)\varepsilon(\lambda_x)\Phi(\lambda_m)D(\lambda_m)c \tag{1}$$

where k is an instrument-dependent constant, I is the excitation light intensity at the sample, ε is the molar absorptivity, Φ is the derivative of the fluorescence quantum yield with respect to emission wavelength, D is the detector response, and c is the concentration (with the optically dilute limit) of the emitting species. The wavelength dependencies are indicated in parentheses.

The extinction coefficients of a molecule, the excitation energy at launch, and the transmission of light in an optical fiber are all wavelength dependent. In addition, the different excitation–emission channels occupy different fiber pairs on the probe button, taking different position on the detection fibers going into the spectrograph and utilizing different cut-off filters at the emission to filter out laser back scattered excitation. All these factors contribute to the different fluorescence responses of the same compound at different excitation–emission channels.

For all LIF measurements, the wavelengths that correspond to the fiber pairs at the probe and to the detection area on the spectrograph/CCD are fixed (or, say, each excitation–emission channel is fixed). The instrument artifacts thus can be corrected. To correct the varying factors that arise from the probe geometry (i.e., the different distance and angle between the fiber pairs due to machining errors) and those from the detection efficiencies of the spectrograph and CCD, a strong beam such as 266 or 355 nm can be used to measure the fluorescence response of a standard compound at each excitation–emission channel. During the measurements, the excitation energies for each measurement are monitored.

B.4.2.2 Energy/Photon Normalization

It has been observed that during a long period of laser operation (days or even months span), laser output might fluctuate greatly. To compare the fluorescence data taken at different periods, it is necessary to normalize all the data to the same excitation energy level. During a short period of time (i.e. $^1/_2$ to 2 hours, which usually covers the period for a LIF/CPT push at a certain location), the energy is fairly stable (Figure B8). Direct measurements of excitation energies are not practical during a LIF/CPT push underground. Therefore, an indirect method to monitor the energy is utilized.

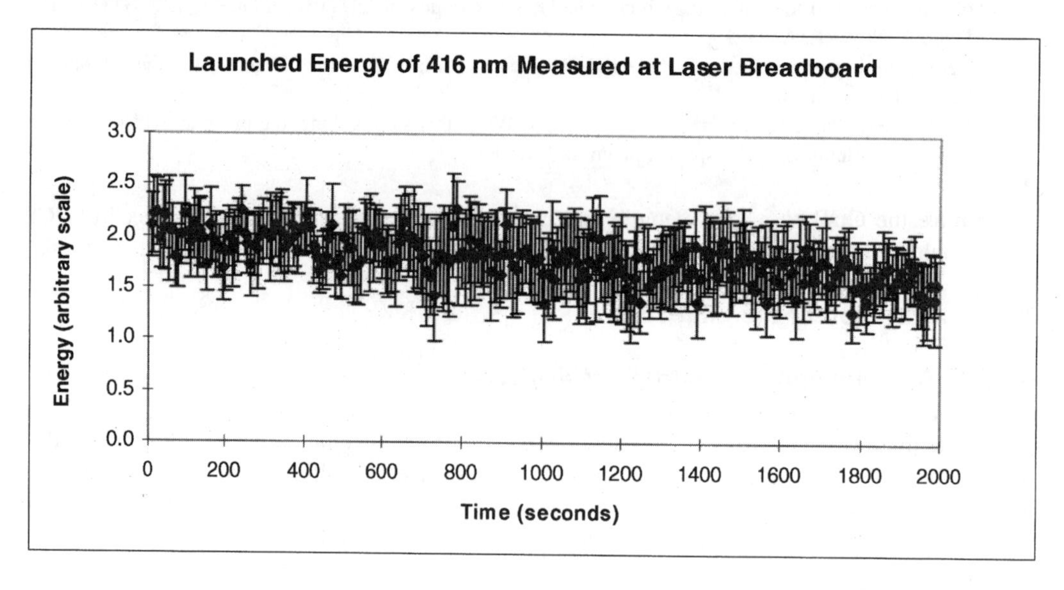

Figure B8. Measurement of 416-nm energy at the laser breadboard over time (2000 seconds).

A standard compound such as quinine sulfate or rhodamine B (RhB) can be used to monitor the fluctuation of excitation energy at each channel by monitoring its fluorescence response, which shows a linear relationship in general. Due to the wide range of excitation wavelengths (from 246.9 to 416 nm) used in the LIF probe, RhB is chosen because it can be excited by all these wavelengths and fluoresce.

The fluorescence vs. excitation energy calibration curves can be obtained with the following procedure:

1. Make a dilute solution of rhodamine B in ethanol (e.g., 0.2 mg/l).
2. Choose the first channel for measurement. Optimize the beam energy at the laser table by centering the fiber at the XY translation stage.
3. Measure the excitation energy at the probe (i.e., in front of the sapphire window) using an energy probe (RjP-627, Laser Probe®) equipped with a radiometer (Rm-6600 Universal, Laser Probe®). Take five measurements.
4. Place the LIF probe (with the sapphire window on the face plate) into the solution. Cover to prevent room light (and other stray light) from being detected.
5. Take spectrum of the RhB solution. Measure five times. Get the average value of its fluorescence peak height (in counts).
6. Take the LIF probe out of the solution. Rinse with ethanol solution and wipe clean.
7. Measure the energy again in front of the probe as stated in step 3. Also take five measurements.
8. Calculate the average of the energies measured at step 3 and step 7. The resulting value is the excitation energy corresponding to the averaged fluorescence peak count taken at step 5.
9. On the laser table, adjust the fiber position at the XY translation stage and move it away from the center of the beam spot to get a less optimized launch energy into the excitation fiber.
10. Redo steps 3 to 8 to get the second sets of peak fluorescence vs. energy.
11. Redo step 9 and 10 to obtain two more sets of peak fluorescence vs. energy in lower values. The energy levels at steps 3 to 11 should cover a range as large as possible.
12. Block the excitation beam at the laser table (or disconnect the connector of the excitation fiber at the patch panel), and redo steps 3 to 8, which will give no fluorescence response (the "peak fluorescence" at this point is merely the baseline of the measurement). This gives the "zero point" of the calibration curve, which shows that when excitation energy is zero no fluorescence is observed.
13. Take the five sets of averaged RhB peak fluorescence counts with their corresponding energies. Fit them to equation $Y = aX$ (where Y is the fluorescence, X is the energy, and a is the linearization coefficient). The calibration curve for this excitation channel is thus completed.
14. Redo steps 2 to 13 for all the other channels within this LIF probe and for all other LIF probes.

To normalize the fluorescence data obtained at different excitation wavelengths, which possess different photon numbers at the same energy level, the calibration needs to go one step further to obtain photon normalization. This can easily be done by converting energies to photons using the following Equations 2 and 3.

$$E_{\text{total}} = \left(10^{-6} tR\right) E_m = \left(10^{-6} tR\right) nh \frac{c}{10^{-9} \lambda} \tag{2}$$

or

$$n = \frac{\left(10^{-6} E_m\right) tR \left(10^{-9} \lambda\right)}{hc} \tag{3}$$

where E_{total} is the total excitation energy, E_m is the pulse energy measured above (it is in μJ per laser pulse), t is the integration time (in seconds) for each measurement, R is the repetition rate of the laser, λ is the wavelength (in nm) of the channel being calibrated, h is the Planck's constant with value of $6.626 \cdot 10^{-34}$ J·S, c is the speed of light with value of $2.998 \cdot 10^{8}$ m/s, and n is the number of photons.

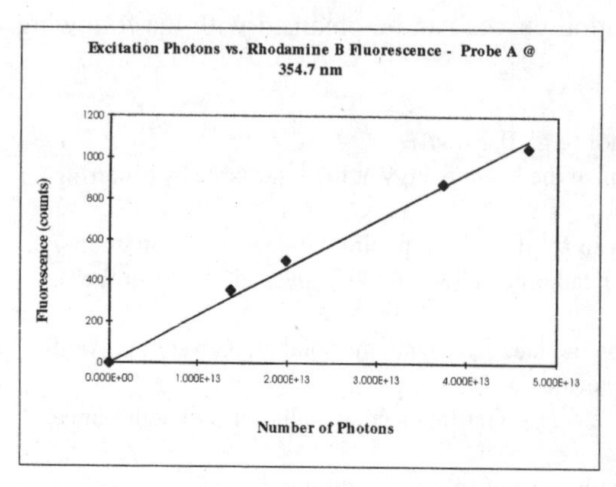

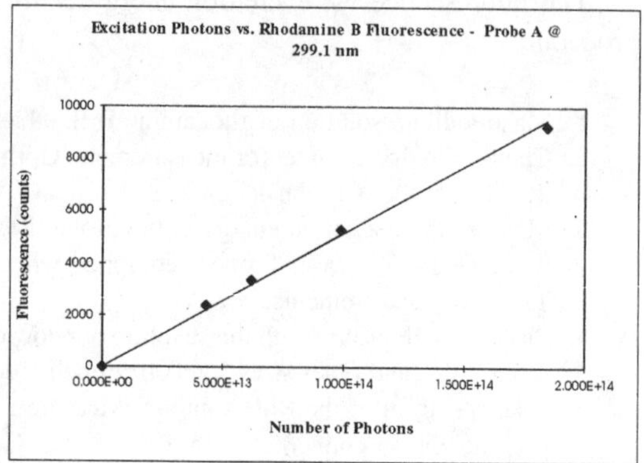

Figure B9. Examples of photon normalization curves.

Figure B9 provides some examples of the calibration curves for the LIF probe. It is observed that the calibration curves achieved good linearization fits with linear-correlation coefficient (R^2) at 0.99 for most of the curves.

By establishing the relationship of RhB fluorescence vs. its excitation energy at each wavelength, we can measure the fluorescence of RhB before and after a LIF measurement (e.g., a LIF/CPT push), take their average, and convert back to the excitation intensity. The linearization coefficient obtained experimentally implicitly includes the effects of molecular response (absorption at the excitation wavelength, fluorescence efficiency), optical fiber attenuation, overlap of the excitation and detection surfaces at the probe/sample interface, and spectrograph imaging. This procedure also allows us to discern the need for laser or fiber adjustment on the excitation before a measurement, depending on fluorescence counts levels seen instantaneously during these calibrations.

B.5 OPERATION OF THE LIF/EEM/CPT TOOL

The operation procedure of the LIF/CPT tool is as follows:

1. Turn on the laser and allow 10 minutes for simmer current to stabilize. When the system is installed in a new environment (e.g., taken from the lab to the field), the simmer current needs to be adjusted to 8.5 V. Allow another 20 minutes for the laser to warm up. An oscilloscope can be used to measure the laser pulse to ensure 50 Hz repetition rate is achieved.
2. While the laser is warming up, the CCD controller is turned on to allow CCD to initialize and to cool down to below 235 K (~15 minutes). All experiment and system calibration should be conducted when the CCD head has cooled down to below 235 K.
3. Exit CCD initialization software and enter the experiment menu. With the software provided, set up the experiment by choosing the correct grating, the slit width, the integration time, CCD detection area, the x-axis for display (i.e., nm or pixel), the center wavelength of the detection, and the number of cycles for accumulation. Save the experiment setup, which can be retrieved later for similar experiments. Note that the experiment can be setup to take numerous spectra automatically and continuously, with all files are saved at the same file directory with an individual file name for each file.
4. When the experiment setup is finished and the system is ready to take the fluorescence spectrum/spectra, enter the file directory and the file name for which the experiment is to be saved. Click Run on the menu and the spectrum/spectra will be taken.
5. The excitation-emission spectra are shown in a three-dimensional display. With a built-in Array Basic® program, the files can be automatically transferred to Microsoft® Excel spreadsheets formats.

The LIF probe assembled to a face plate with a sapphire window can be used for measurements conducted in the lab. Before a field operation, the probe is assembled into the CPT "sub" (Design Manual Chapter III C).

It is conducted using the following steps:

1. The connectors of the probe at the patch panel are disconnected and taped with electric tape into a bundle.
2. Slide this connector end of the probe through the CPT rods, which are later attached to the CPT cone, and the hosting rod for the hydraulic system.
3. Pull the cable straight until the face plate is positioned in front of its housing position on the CPT sub (Design Manual Chapter III C). Fit the face plate onto the sub.
4. Place the depth gauge or other CPT sensing probe with its cable below the LIF probe through the same rods. Connect the sub to the cone.
5. Slide the LIF probe cable through the rod segments until about 5 m of the cable is left.
6. Place the hosting rod of the hydraulic system with the sub housing the LIF probe in position.
7. Take off the electric tape on the connectors of the probe. Connect them to the patch panel.
8. Place filters for the necessary channels between the connectors that connect emission fibers at the patch panel.
9. When the laser and CCD are stabilized, the system is ready for LIF/CPT measurement.

When the CPT cone is advancing, there are two approaches for data acquisition. First is the stop-run-stop approach; when the CPT cone advances to certain depth the operator runs the data acquisition software at GRAMS/386®. This routine is more time consuming. The other approach is to set up the software to external triggering. When this routine is used, the computer accepts pulse signals sent by the CPT depth gauge and in turn triggers the data acquisition routine and saves the spectra taken at each depth.

Five fluorescence spectra of RhB stock solution are taken before and after each push (after decontamination process). The average of the fluorescence signal is converted back to photons using the calibration curve obtained earlier. Data taken at different locations are normalized to photons.

When the laser and CCD are powered by the internal generator at the CPT vehicle, the laser and the CCD have to be shut down after each push while the vehicle is moving to a new location and then be turned on again before any new measurements can be taken. This slows down the productivity of the LIF-CPT operation. To avoid this problem, an external generator can be used to continuously provide power. If the generator output is noisy, an inverter can be used to "clean" the generator output.

GLOSSARY OF TERMS

AATDF: U.S. Department of Defense Advanced Applied Technology Demonstration Facility at the Energy and Environmental Systems Institutes of Rice University
ANGB: U.S. Air National Guard Base
BTEX: Benzene, toluene (methylbenzene), ethylbenzene, and xylene (dimethylbenzene; includes three structural isomers: *o*-, *m*-, *p*-xylene)
CCD: Charge-couple detector
CPT: Cone penetrometer technology
EEM: Excitation–emission matrix
GC/MS: Gas chromatography/mass spectrometry
HG: Harmonic generator
HPLC: High-performance liquid chromatography
LIF: Laser-induced fluorescence

NA: Numerical aperture
PAH: Polynuclear aromatic hydrocarbons (or polyaromatic hydrocarbons)
ROST: Rapid Optical Screening Tool
SCAPS: Tri-Service Characterization and Analysis Penetrometer System
SRS: Stimulated Raman Scattering

REFERENCES

1. S.J. Hart, Y.-M. Chen, B.K. Lien, J.E. Kenny. 1996. "A fiber optic multichannel laser spectrometer system for remote fluorescence detection in soils" *SPIE*, August, Denver, Colorado.
2. J. Lin, S.J. Hart, W. Wang, D. Namytchkine, J.E. Kenny. 1995. "Subsurface Contaminant Monitoring by Laser Fluorescence Excitation-Emission Spectroscopy in a Cone Penetrometer Probe," *European Symposium on Optics for Environmental and Public Safety* 19–23 June, Munich, Germany.
3. J. Lin, S.J. Hart, T.A. Taylor, J.E. Kenny. 1994. "Laser fluorescence EEM probe for cone penetrometer pollution analysis" *A&WMA/SPIE Symposium on Optical Sensing for Environmental and Process Monitoring*, 2367, 70–79.
4. S.H. Lieberman, G.A. Theriault, S.S. Cooper, P.G. Malone, R.S. Olsen, P.W. Lurk. 1991. "Rapid, subsurface, in-situ field screening of petroleum hydrocarbon contamination using laser induced fluorescence over optical fibers" *Proc. Second International Symposium, Field Screening Methods for Hazardous Waste Site Investigations*, 57–63.
5. R.W. St. Germain, G.D. Gillispie. 1995. "Real-time continuous measurement of subsurface petroleum contamination with the Rapid Optical Screening Tool (ROST™)," *Proc. U.S. EPA and A&WMA International Symposium on Field Screening Methods for Hazardous Wastes and Toxic Chemicals,* 467–77.
6. G.B. Jarvis, S. Mathew, J.E. Kenny. 1994. "Evaluation of Nd:YAG-pumped Raman shifter as a broad-spectrum light source" *Applied Optics*, 33(21), 4938–4946.
7. S. Mathew. 1997. "Fluorescence excitation-emission matrix measurement and analysis of ground water contaminants," *A Dissertation for the Degree of Doctor of Philosophy in Chemistry,* Tufts University, September.
8. J. Lin, S.J. Hart, J.E. Kenny. 1996. "Improved two-fiber probe for in situ spectroscopic measurement" *Anal. Chem.*, 68(18), 3098–3103.
9. Augat Communication Products, Inc. 1993. "Fiber optics assembly instructions, ST connectors, 490196" April, Rev B.
10. S.A. Quantel. 1996. "Quantel brilliant Q-switched Nd:YAG laser instruction manual."
11. Instrument SA, Inc. 1995. Jobin-Yvon – SPEX "Manual — Spectrum one CCD detection system" part number 80119 Rev F, Revised October.
12. J.V. Sweedler. 1993. "Charge transfer device detectors and their applications to chemical analysis" *Crit. Rev. Anal. Chem.*, 24(1), 59–98.
13. Instrument SA, Inc. 1997. Jobin-Yvon – SPEX "Data collection handbook — SpectraMax Version 2.0" Part Number 81006, May.

APPENDIX C

Excitation–Emission Matrices Collected in the Laboratory, from Hanscom AFB and Otis ANGB

P-Xylene
Anthracene
Chrysene
Fluoranthene
Naphthalene
2-methylnaphthalene
Naphthalene and 2-methylnaphthalene mixture
Naphthalene, 2-methylnaphthalene, and phenanthrene mixture
Naphthalene and anthracene mixture
Naphthalene, anthracene, and fluoranthene mixture
P-Xylene, naphthalene, anthracene, and fluoranthene mixture
Phenanthrene
Phenanthrene and pyrene mixture
Phenanthrene, pyrene, and benzo(b)fluoranthene mixture
Phenanthrene, pyrene, benzo(a)anthracene and benzo(b)fluoranthene mixture
Phenanthrene, pyrene, benzo(a)anthracene, benzo(b)fluoranthene and P-xylene mixture
Pyrene
Benzo(a)pyrene
Benzo(a)anthracene
Benzo(b)fluoranthene
Hanford AFB: HAFB well MWZ-9 sample
HAFB well MWZ-18 sample
HAFB well MWZ-20 sample
Otis ANGB: OANGB CY4J 234cm
OANGB CY4J 314cm
OANGB CY4J 652cm
OANGB CY4K 174cm
OANGB CY4M 190cm
OANGB CY4P 148cm
OANGB CY4P 258cm
OANGB CY4P 714cm
OANGB CY4Q 214cm
OANGB CY4Q 672cm

Gasoline
Heating oil
Creosote in water and diesel in water
JP-4 and JP-8

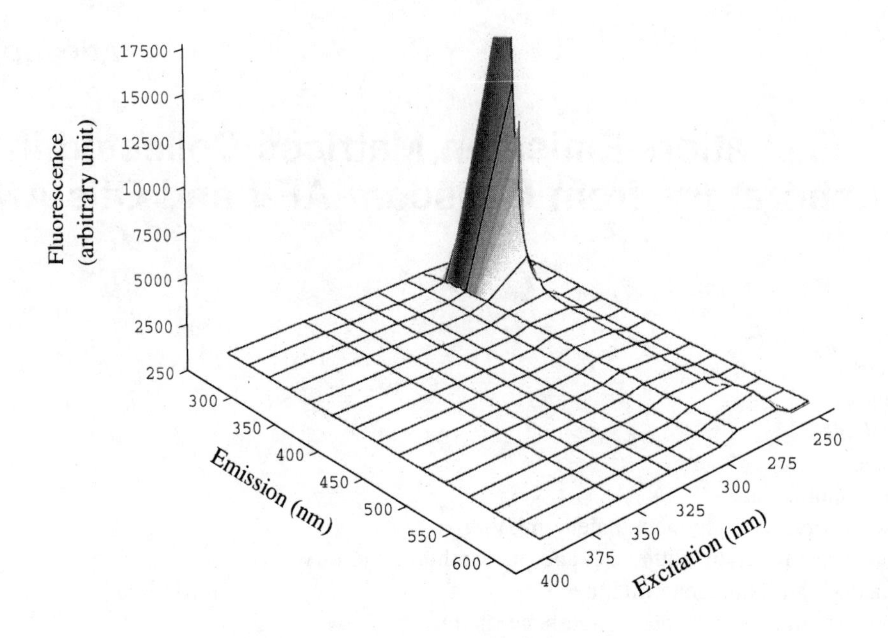

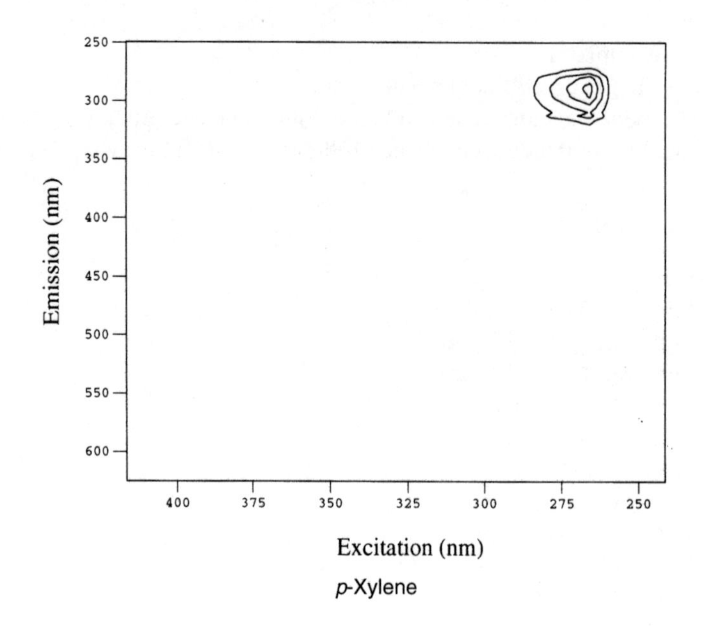

Excitation (nm)

p-Xylene

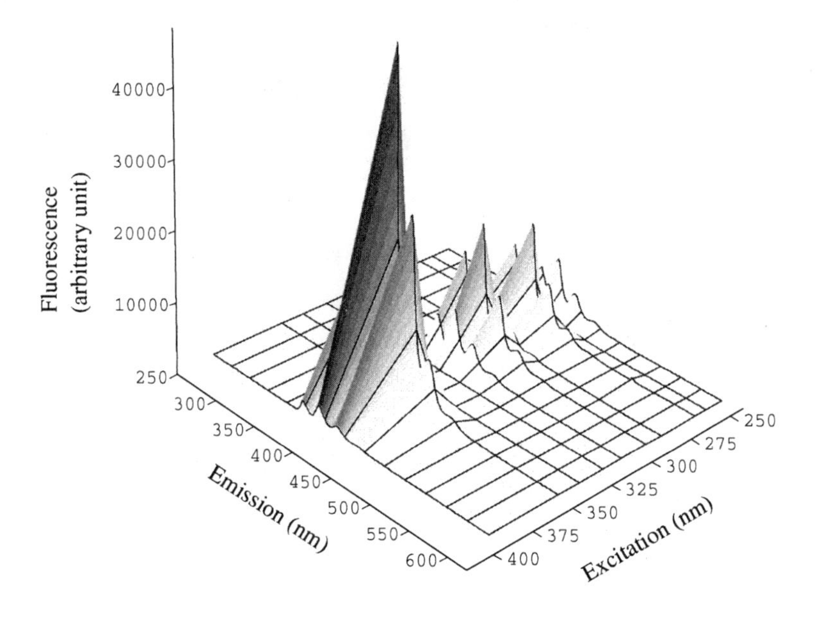
Fluorescence (arbitrary unit)
40000
30000
20000
10000
250
300
350
400
450
500
550
600
Emission (nm)
250
275
300
325
350
375
400
Excitation (nm)

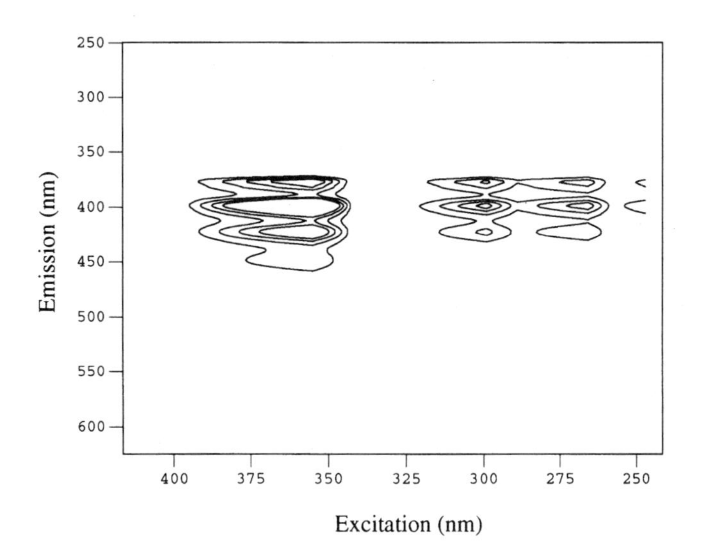
250
300
350
400
450
500
550
600
Emission (nm)
400
375
350
325
300
275
250
Excitation (nm)

Anthracene

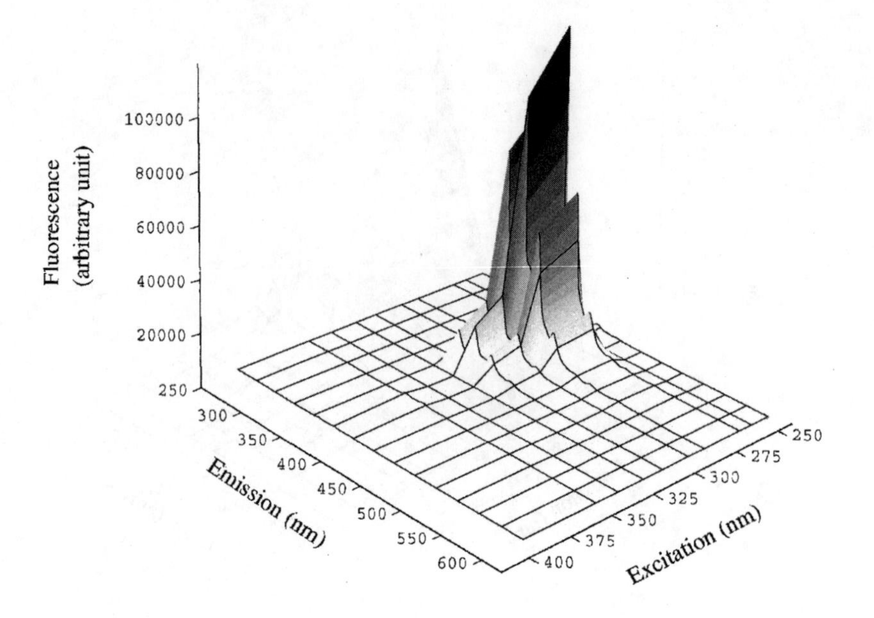
Fluorescence
(arbitrary unit)
100000
80000
60000
40000
20000
250
300
350
400
450
500
550
600
Emission (nm)
250
275
300
325
350
375
400
Excitation (nm)

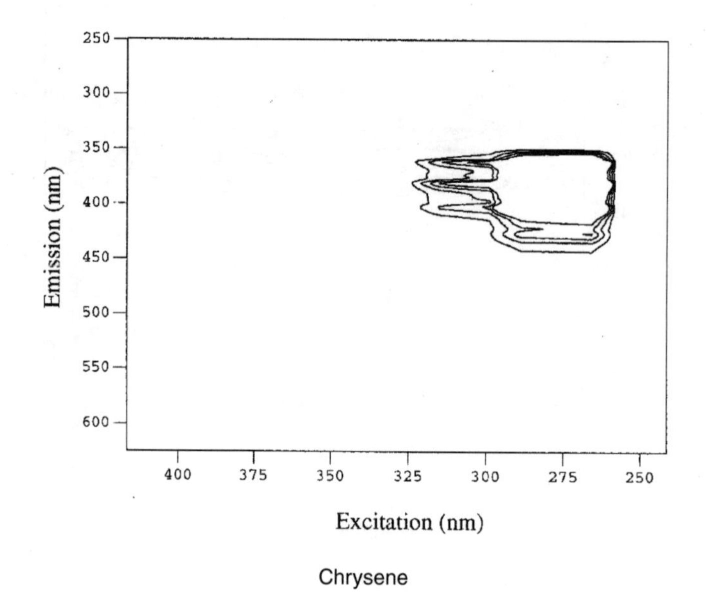
Emission (nm)
250
300
350
400
450
500
550
600
400
375
350
325
300
275
250
Excitation (nm)

Chrysene

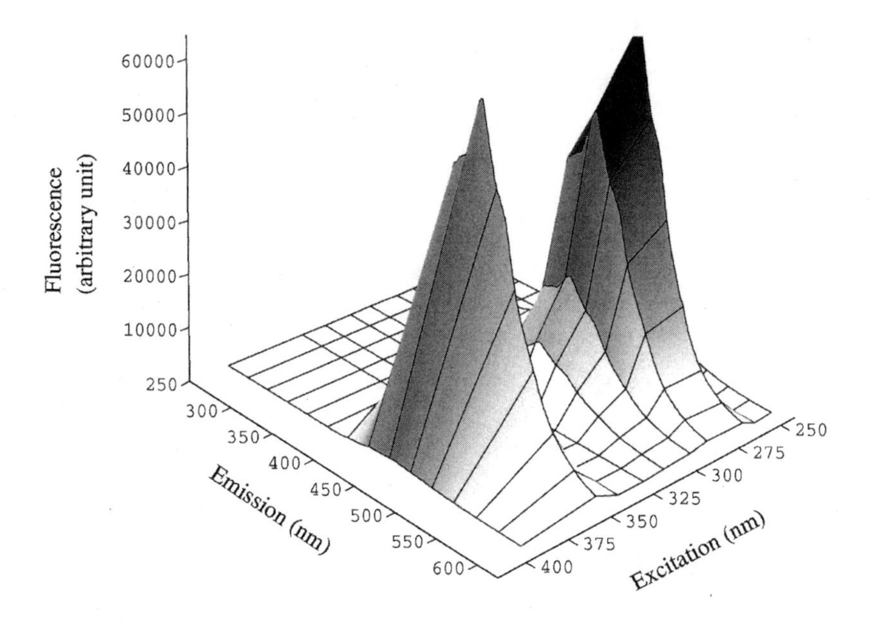

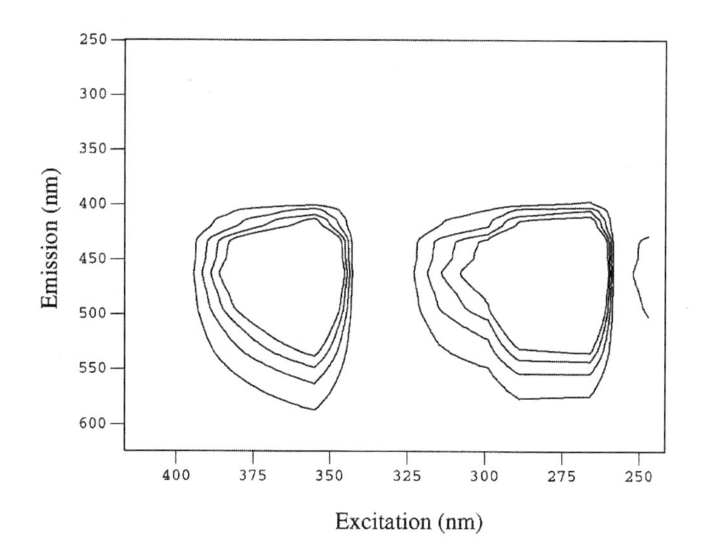

Fluoranthene

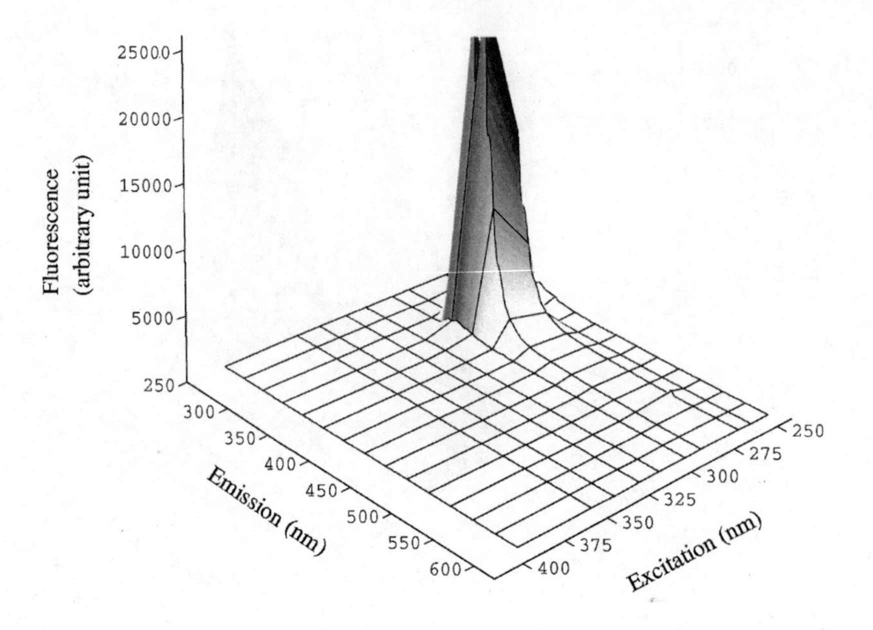

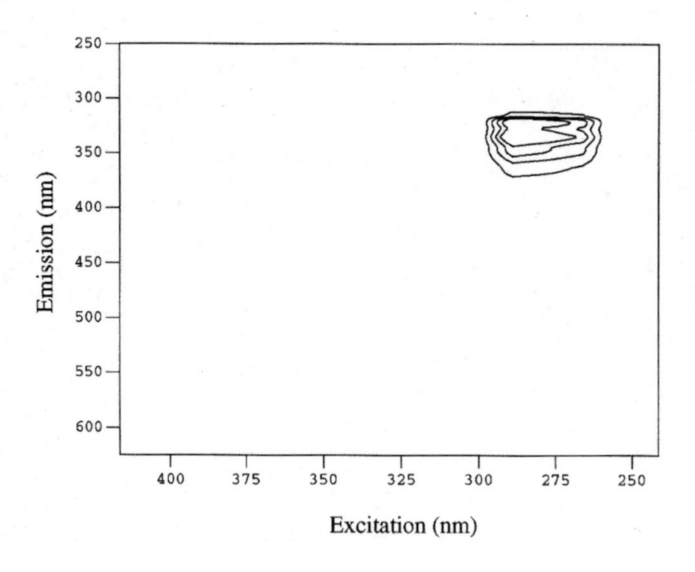

Naphthalene

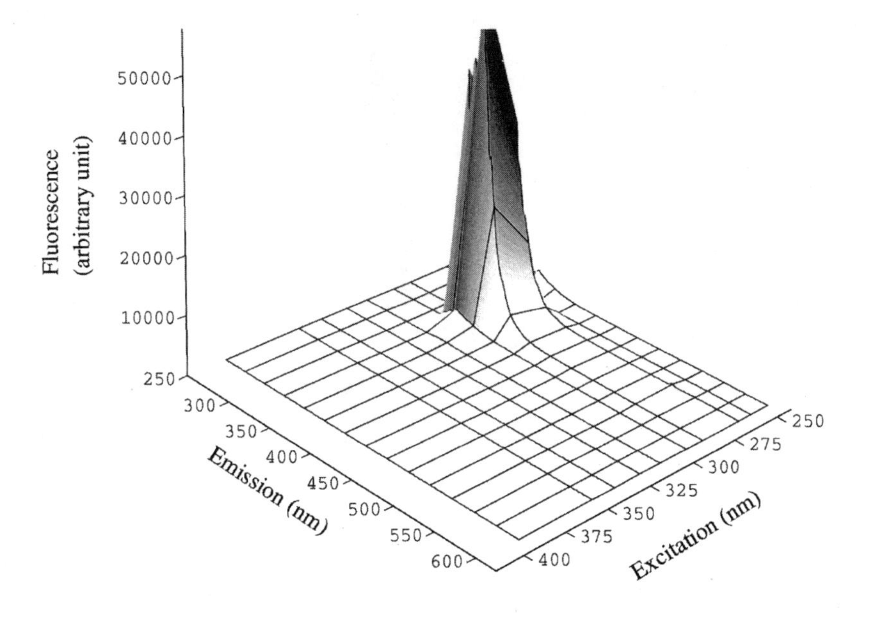

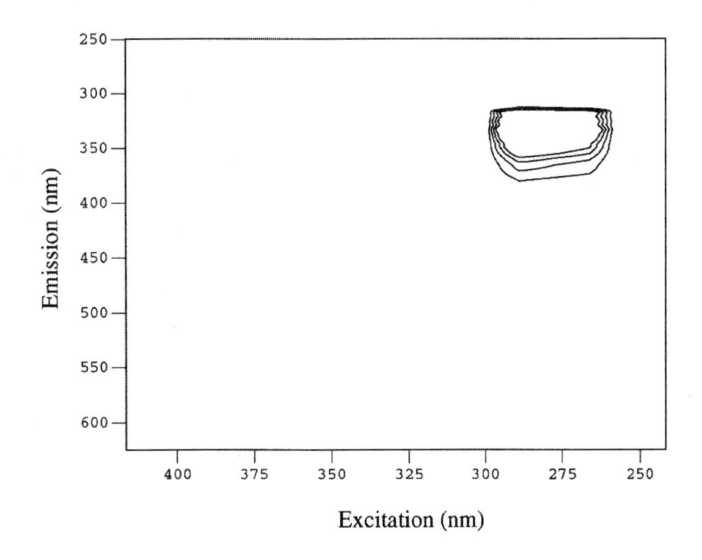

2-methylnaphthalene

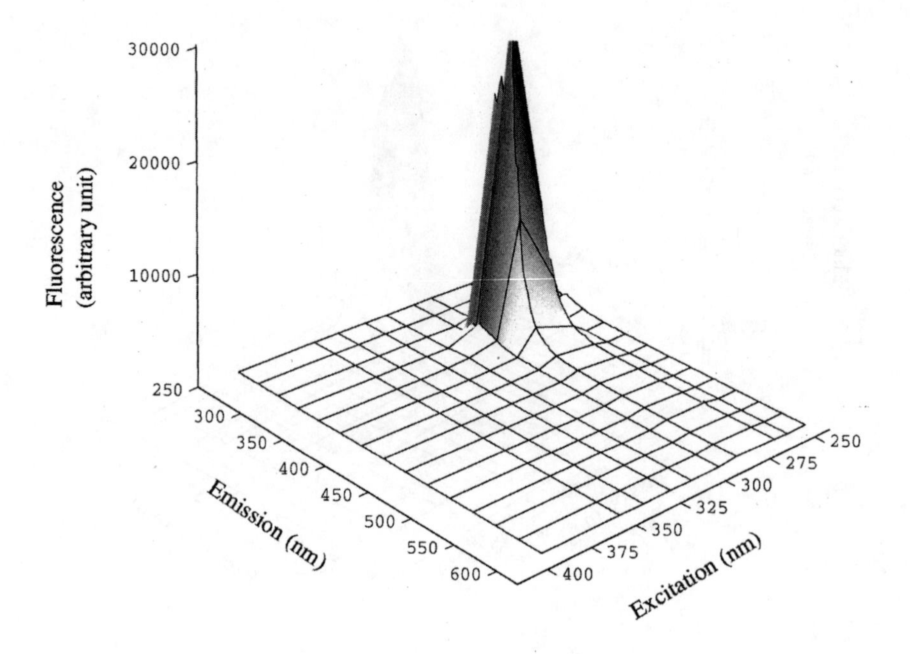

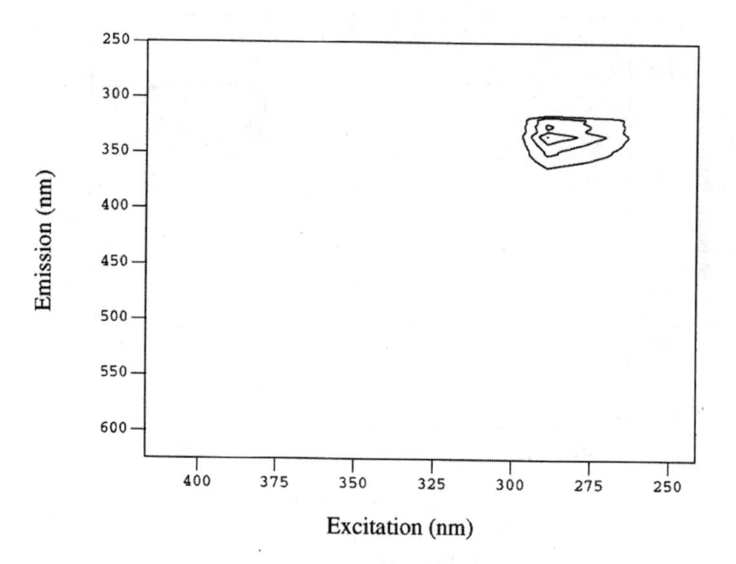

Naphthalene and 2-methylnaphthalene mixture

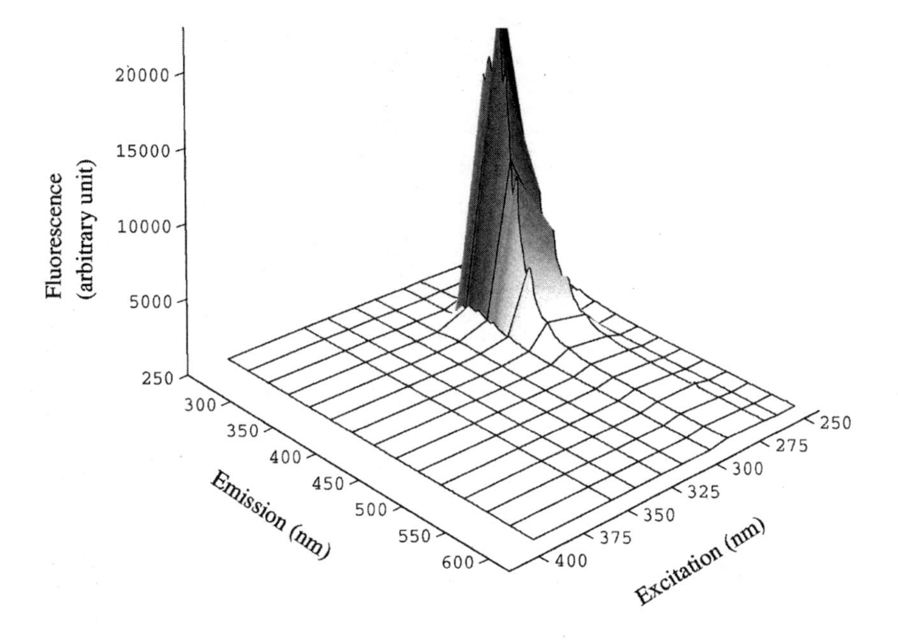

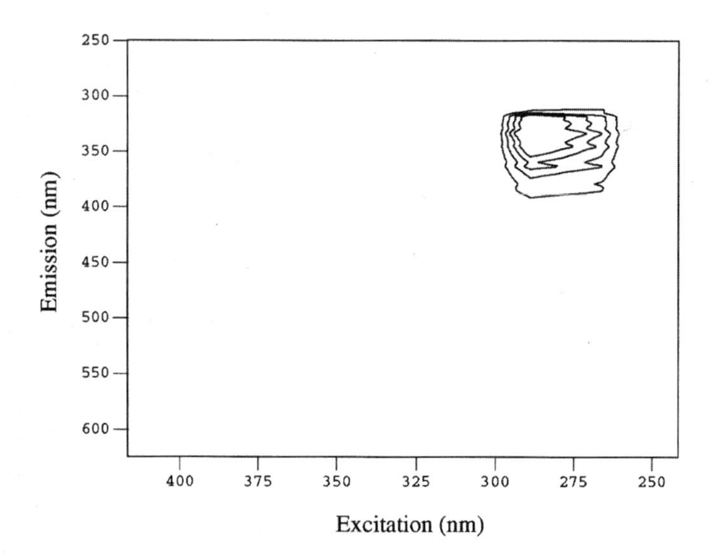

Naphthalene, 2-methylnaphthalene, and phenanthrene mixture

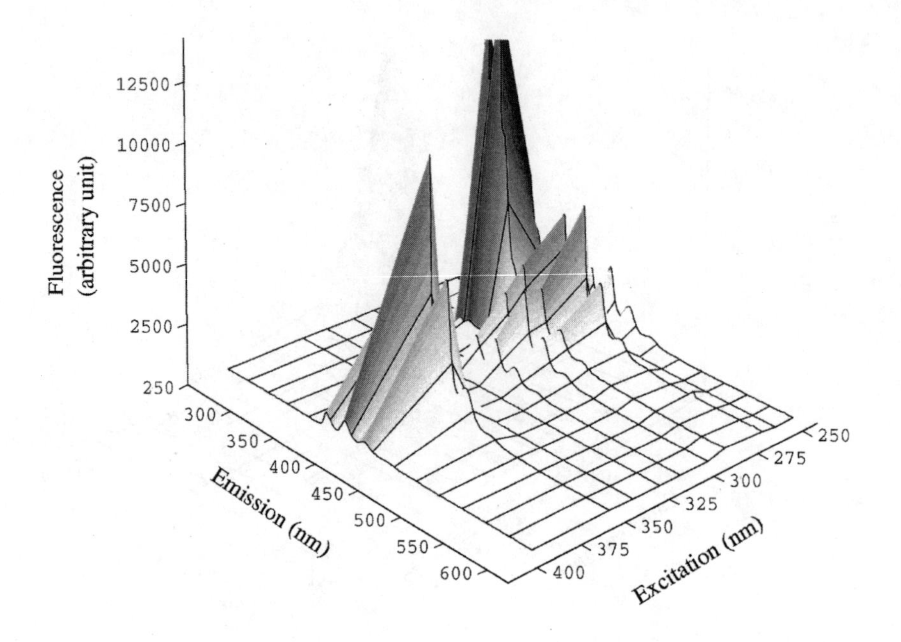

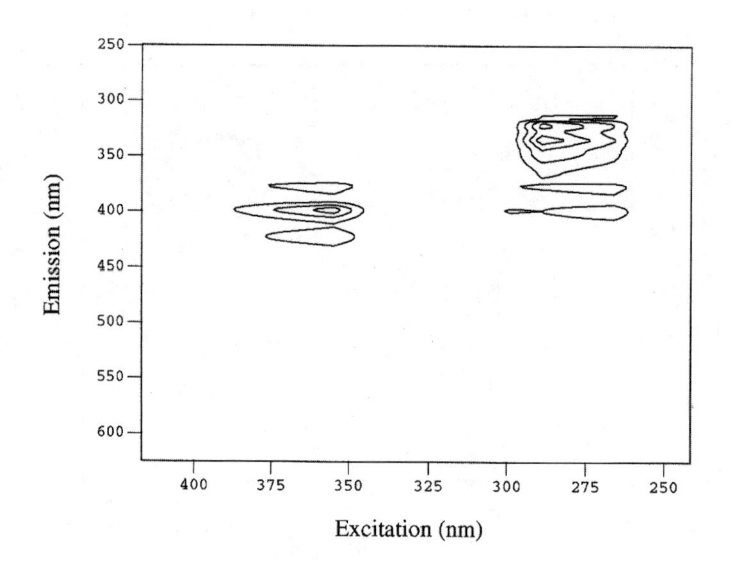

Naphthalene and anthracene mixture

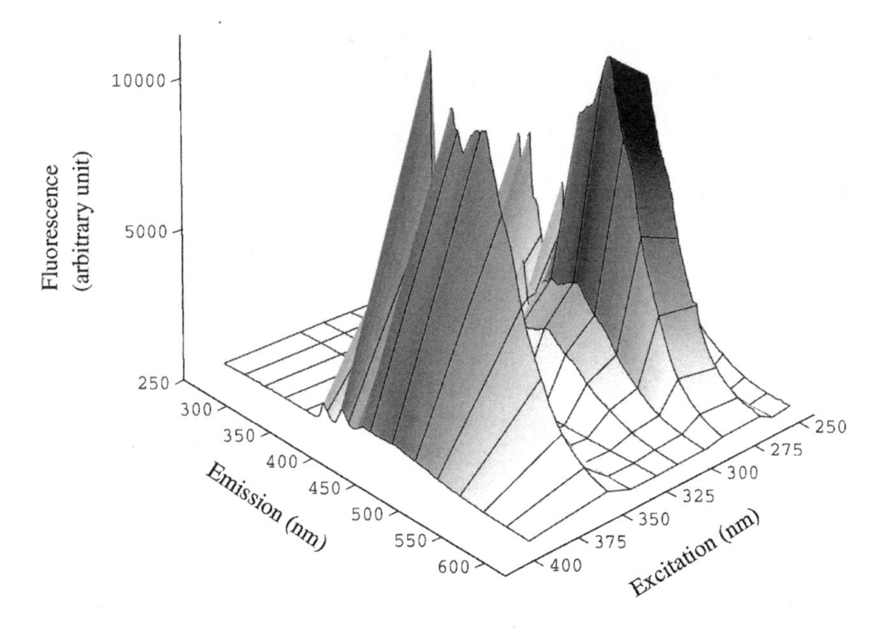

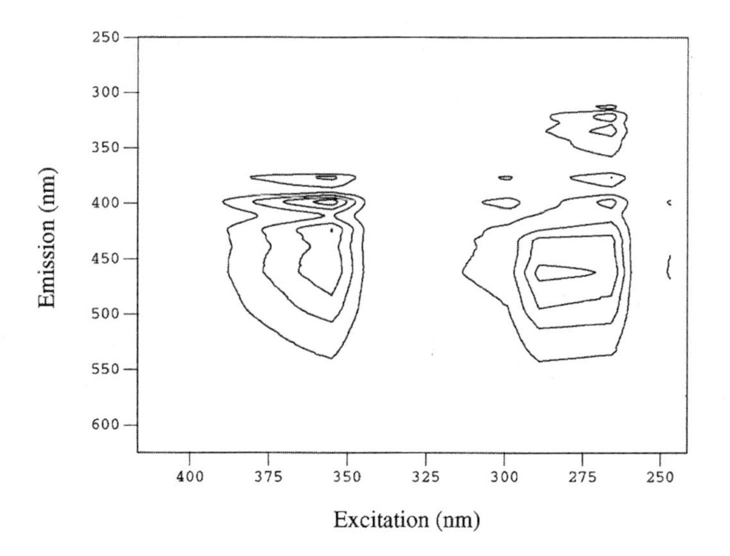

Naphthalene, anthracene, and fluoranthene mixture

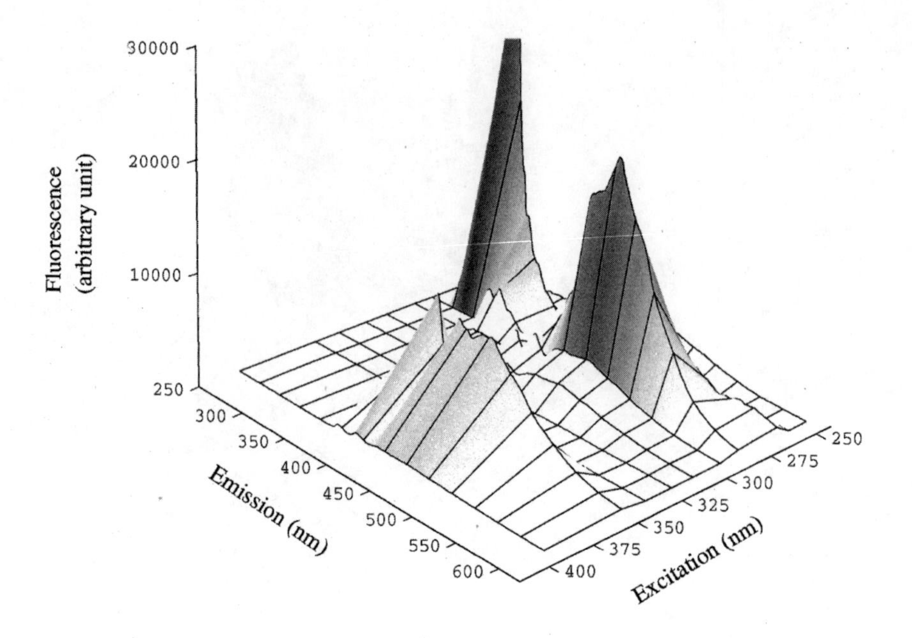

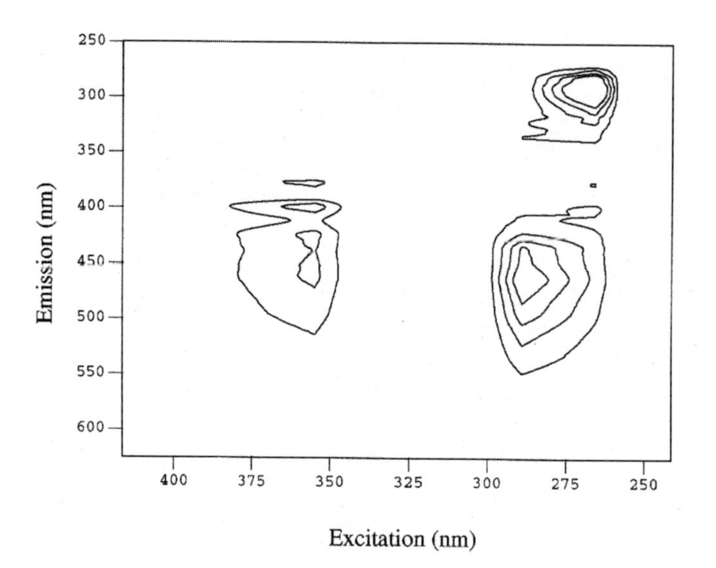

p-Xylene, naphthalene, anthracene, and fluoranthene mixture

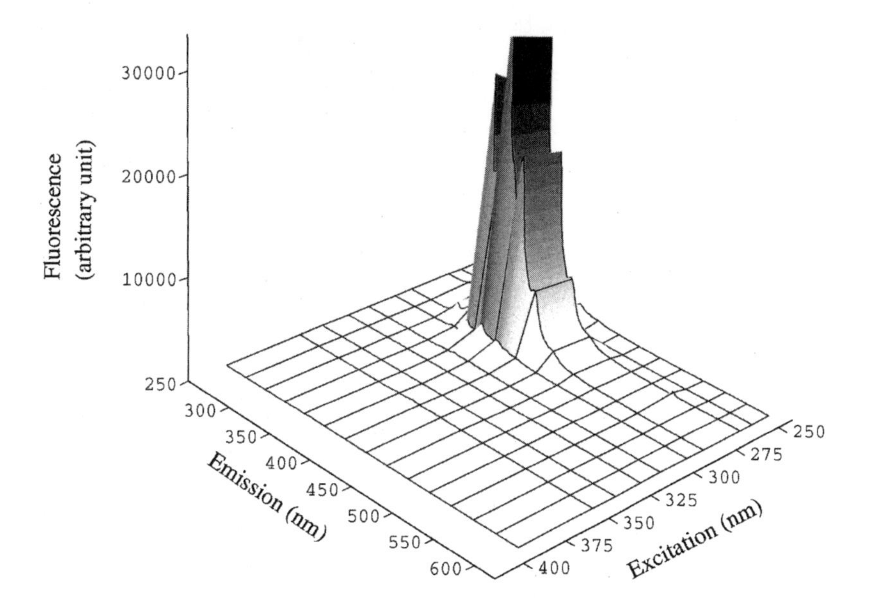

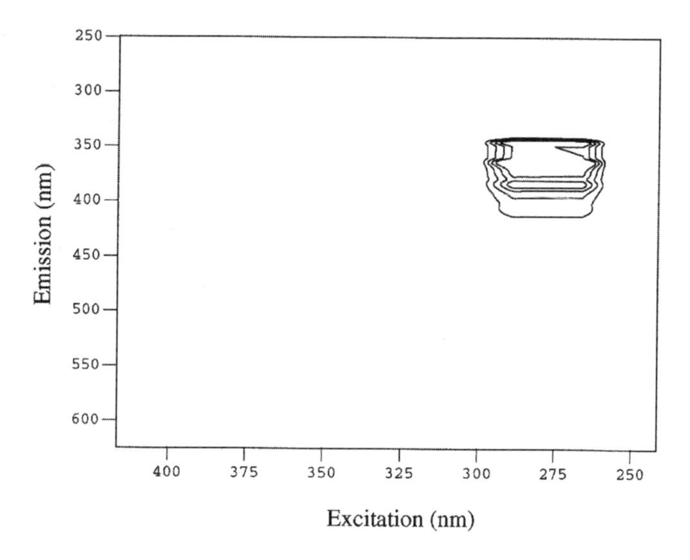

Phenanthrene

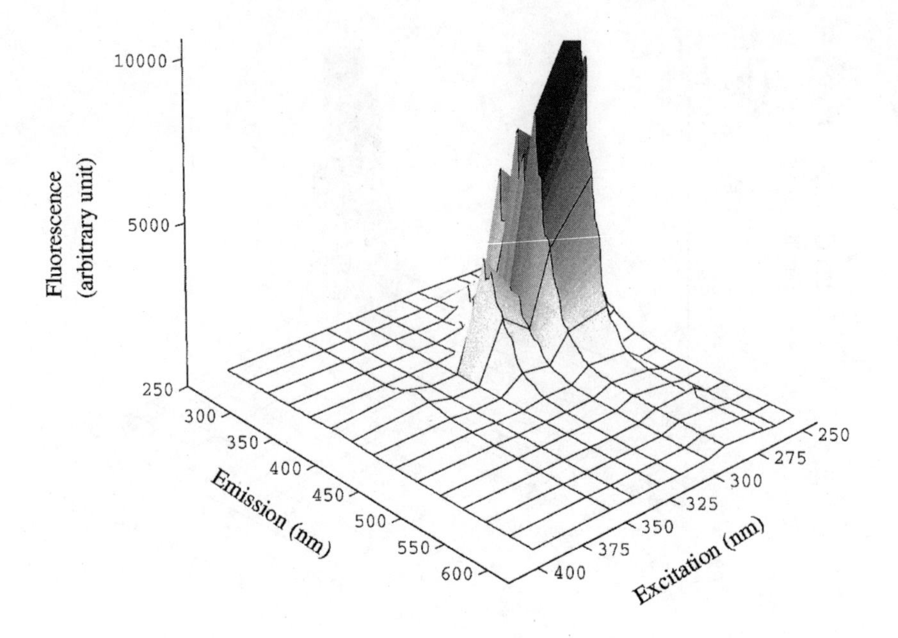

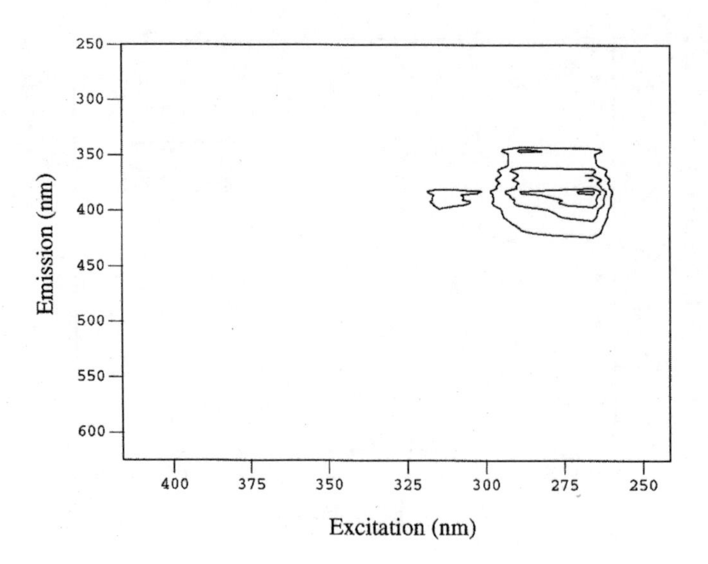

Phenanthrene and pyrene mixture

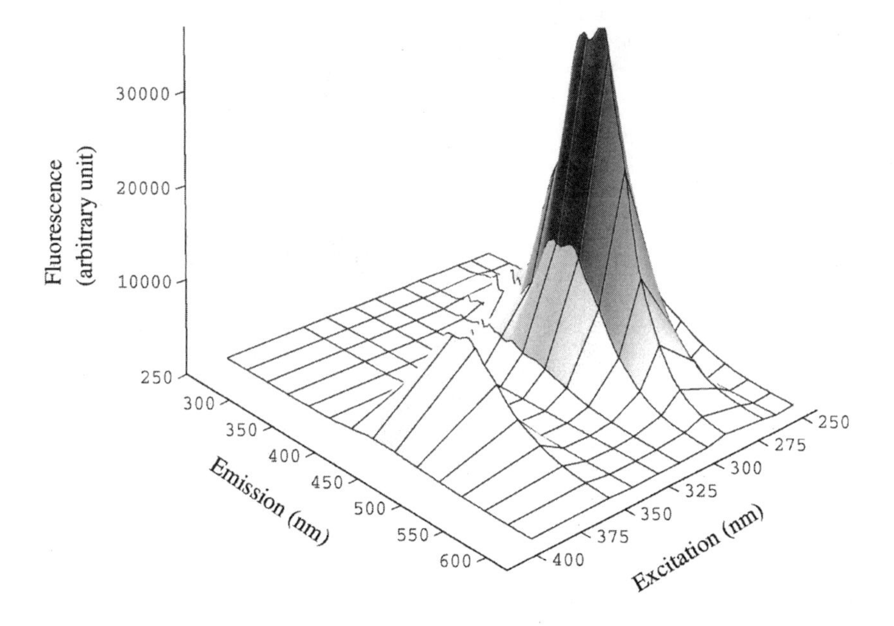

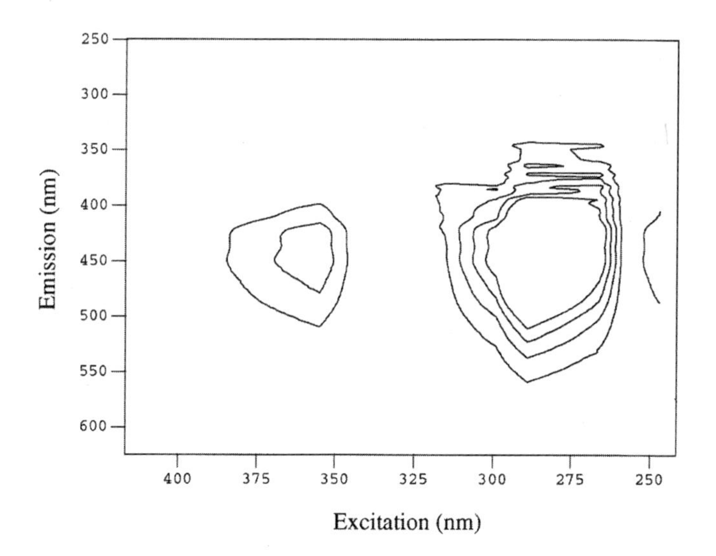

Phenanthrene, pyrene, and benzo(b)fluoranthene mixture

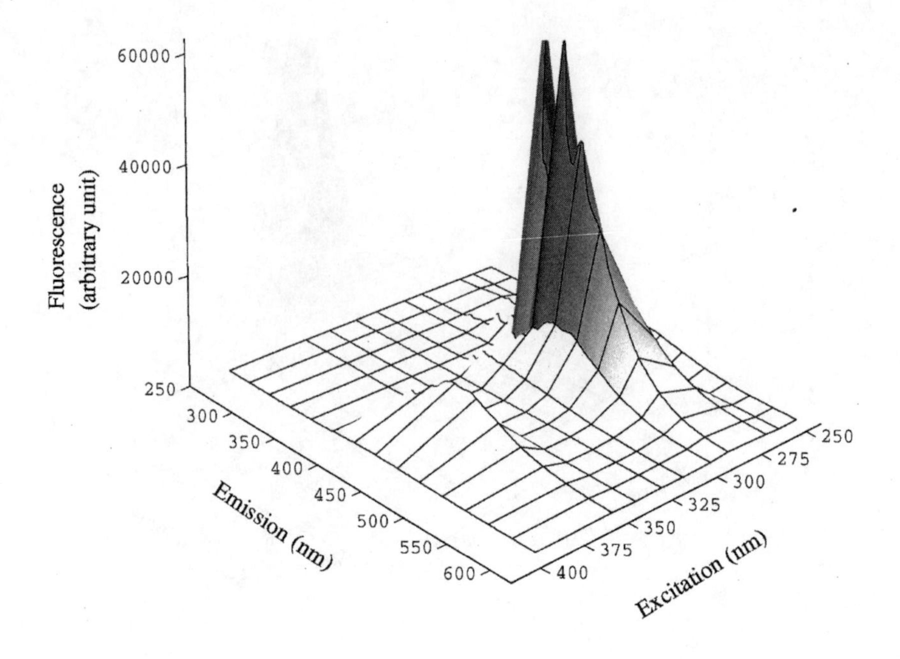

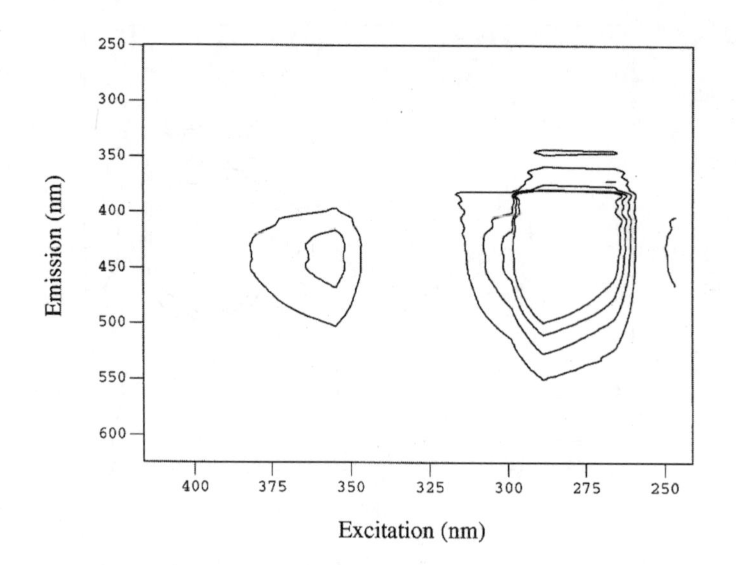

Phenanthrene, pyrene, benzo(a)anthracene, and benzo(b)fluoranthene mixture

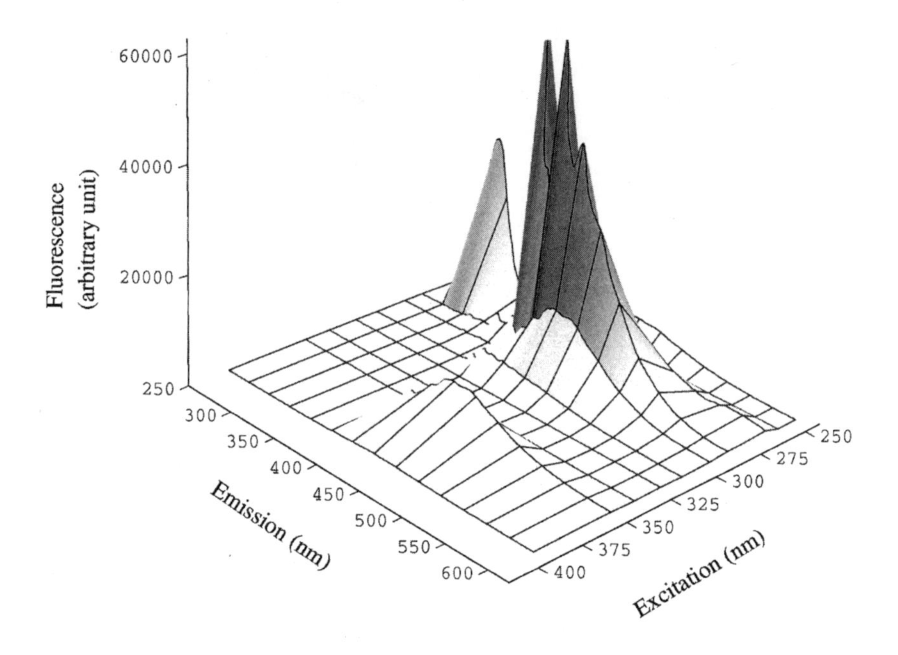

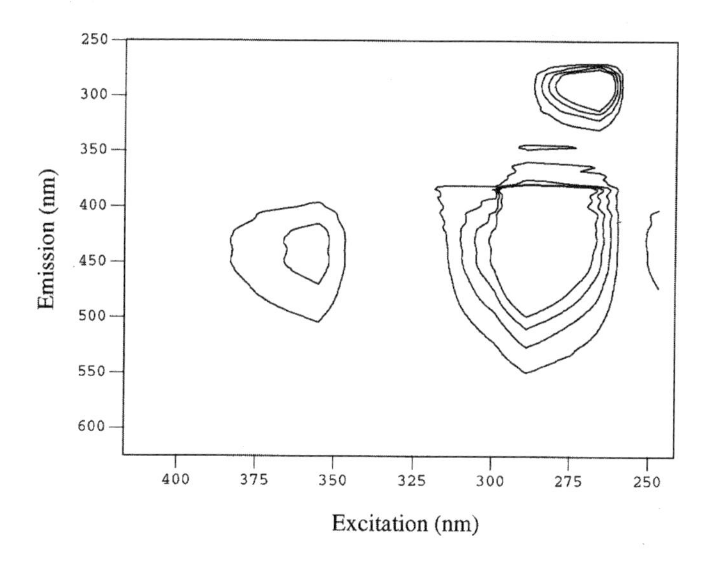

Phenanthrene, pyrene, benzo(a)anthracene, benzo(b)fluoranthene, and *p*-xylene mixture

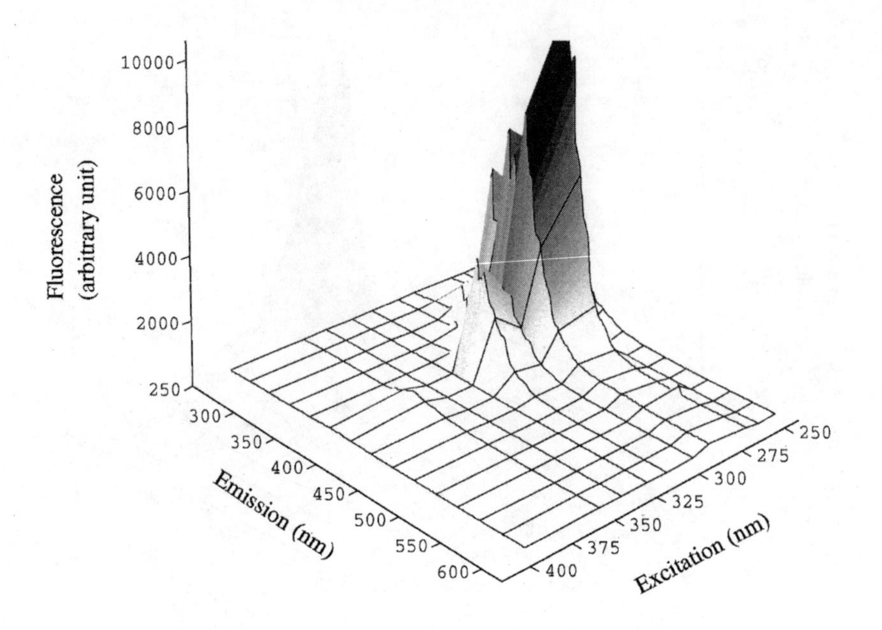

Fluorescence (arbitrary unit)
10000
8000
6000
4000
2000
250
300
350
400
450
500
550
600
Emission (nm)
250
275
300
325
350
375
400
Excitation (nm)

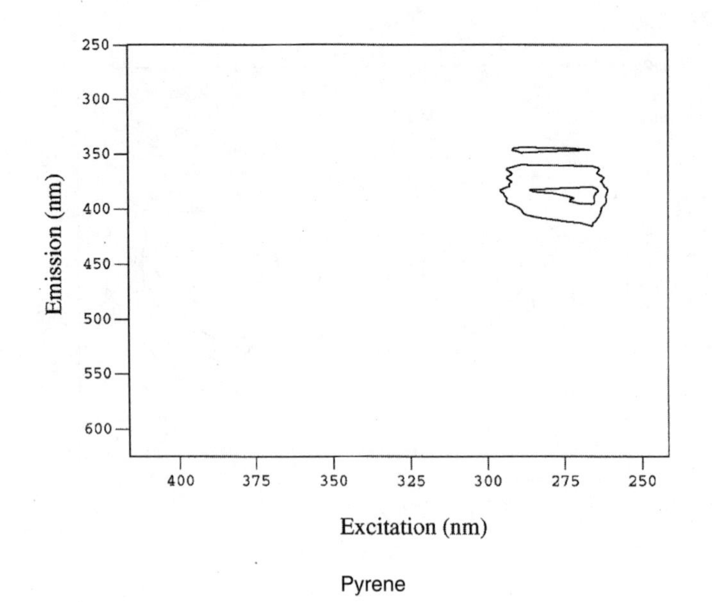

Emission (nm)
250
300
350
400
450
500
550
600
400
375
350
325
300
275
250
Excitation (nm)

Pyrene

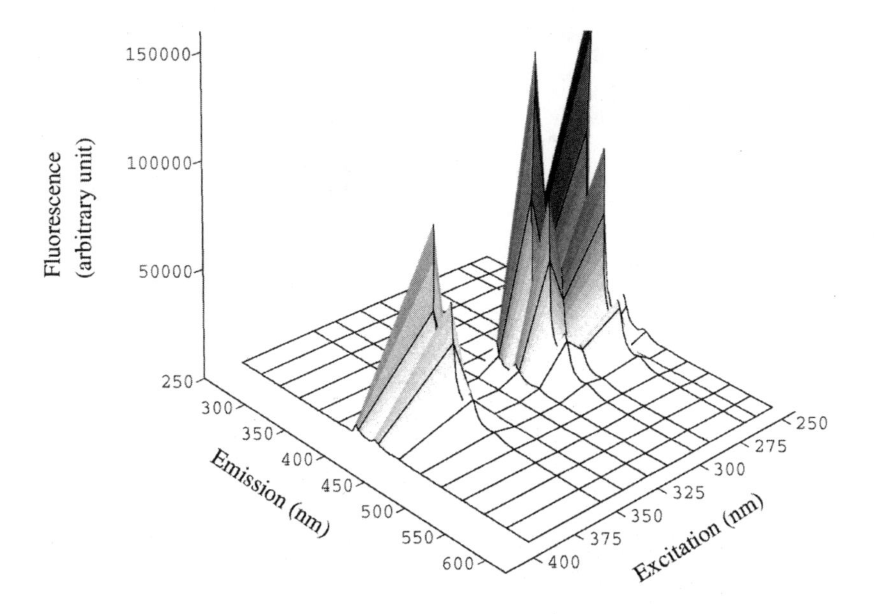

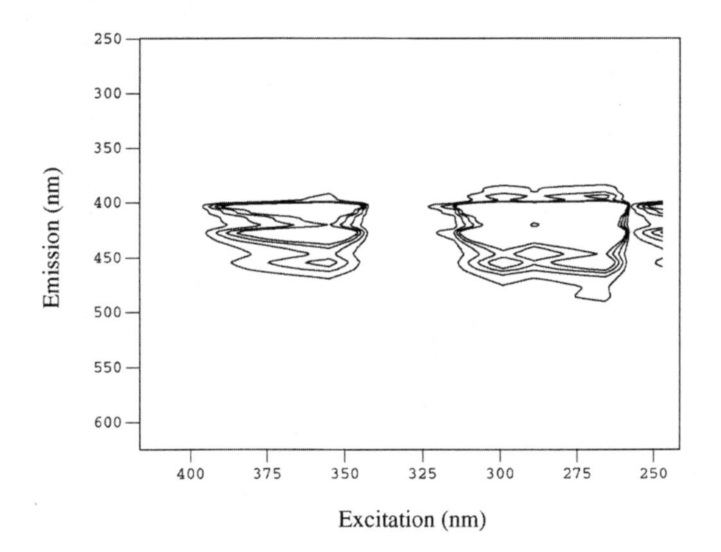

Benzo(a)pyrene

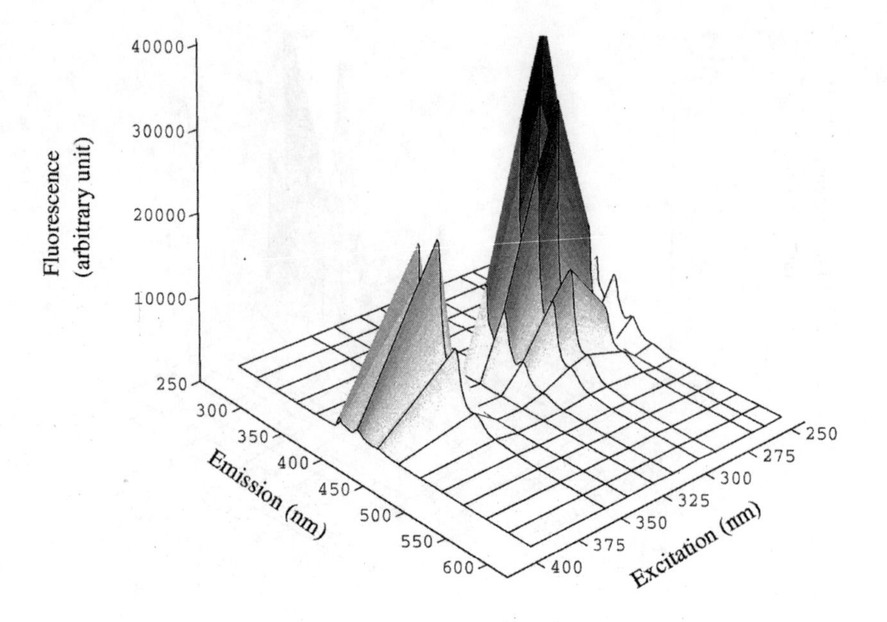

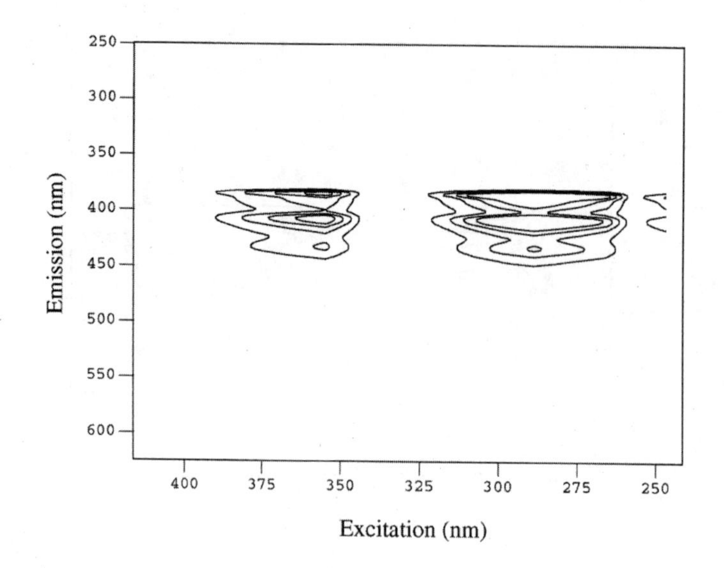

Benzo(a)anthracene

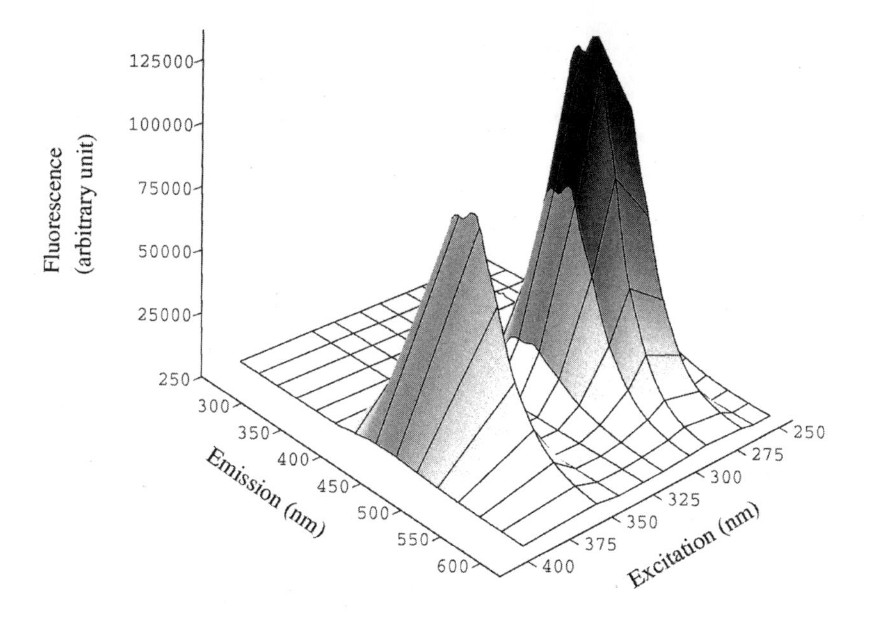

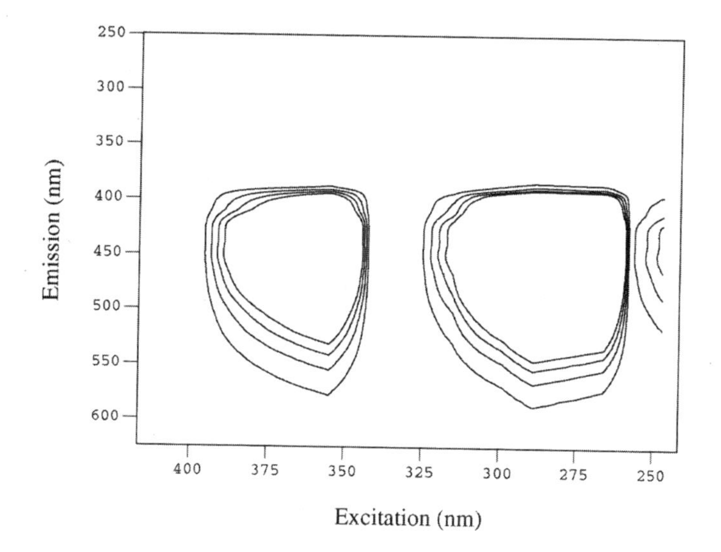

Benzo(b)fluoranthene

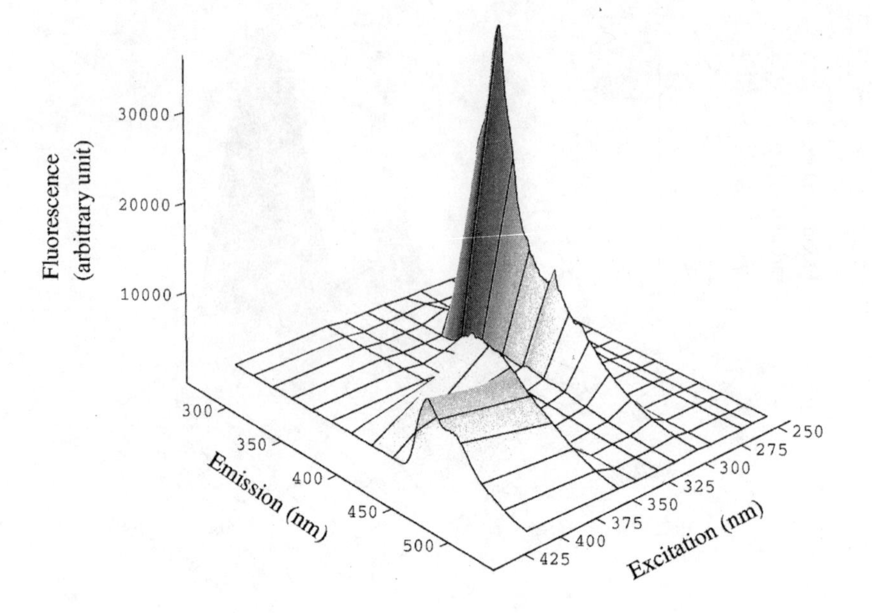

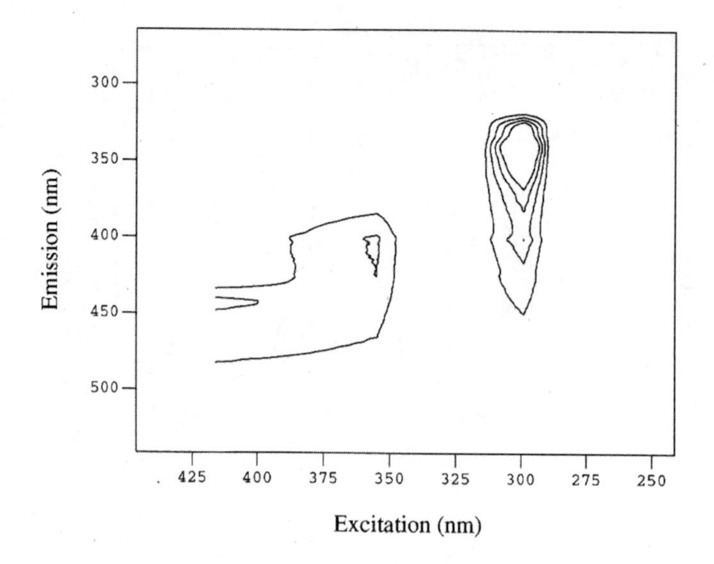

HAFB well MWZ-9 sample

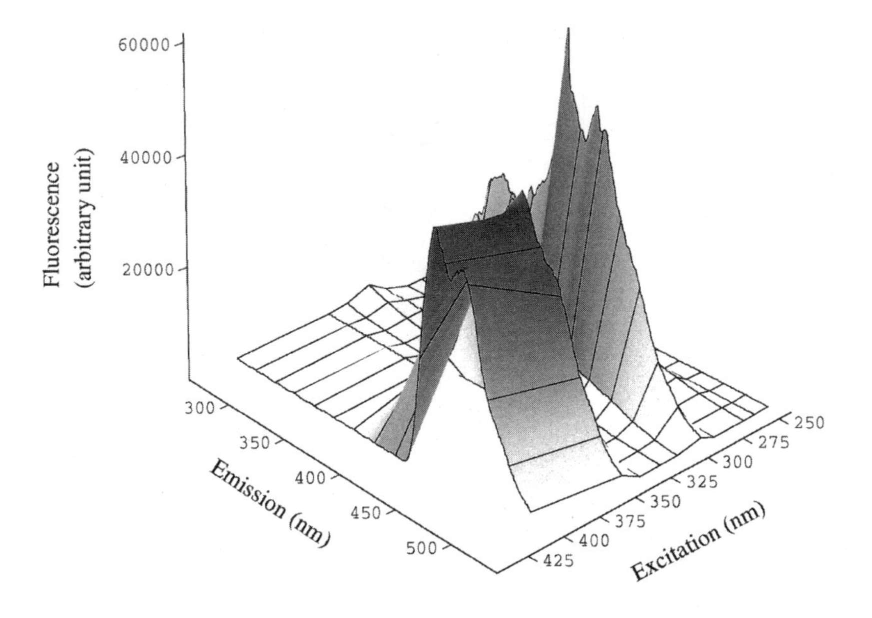

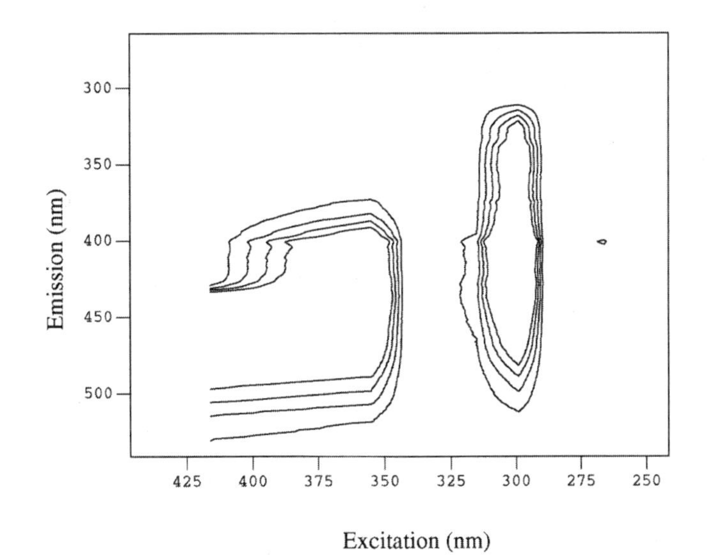

HAFB well MWZ-18 sample

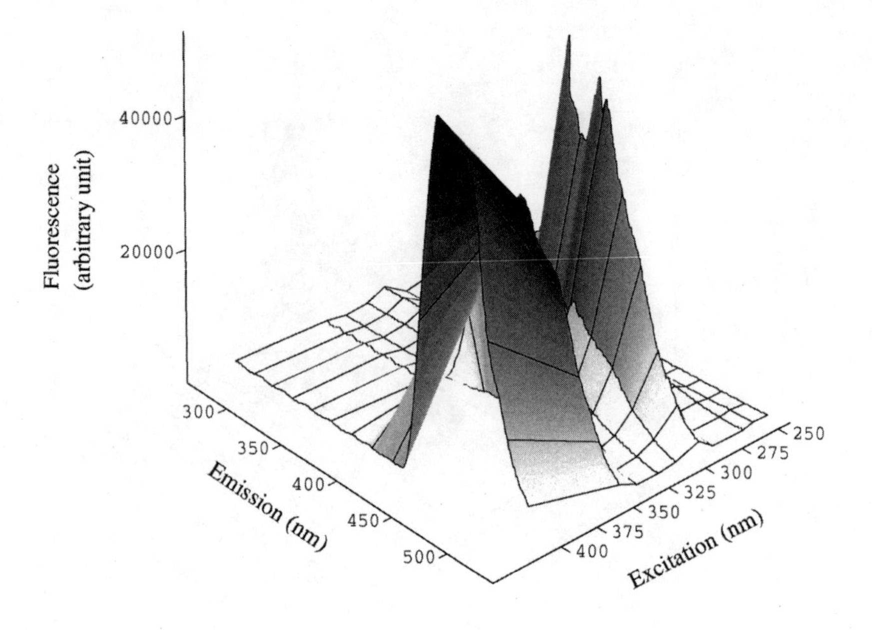

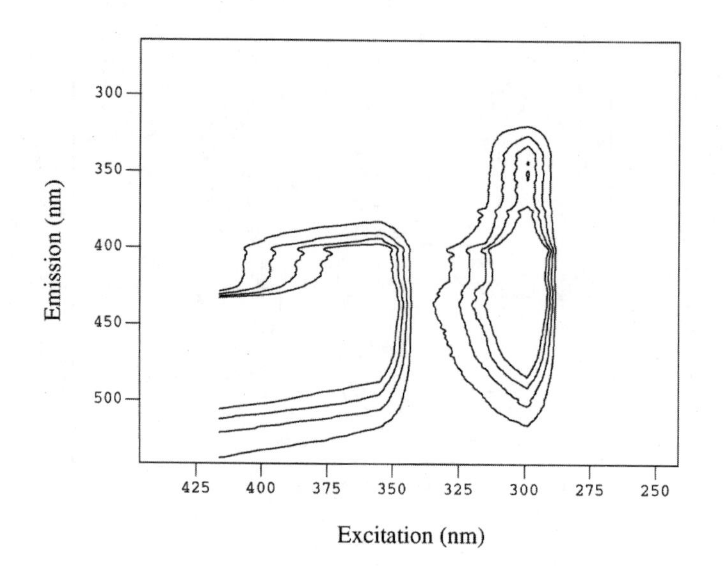

HAFB well MWZ-20 sample

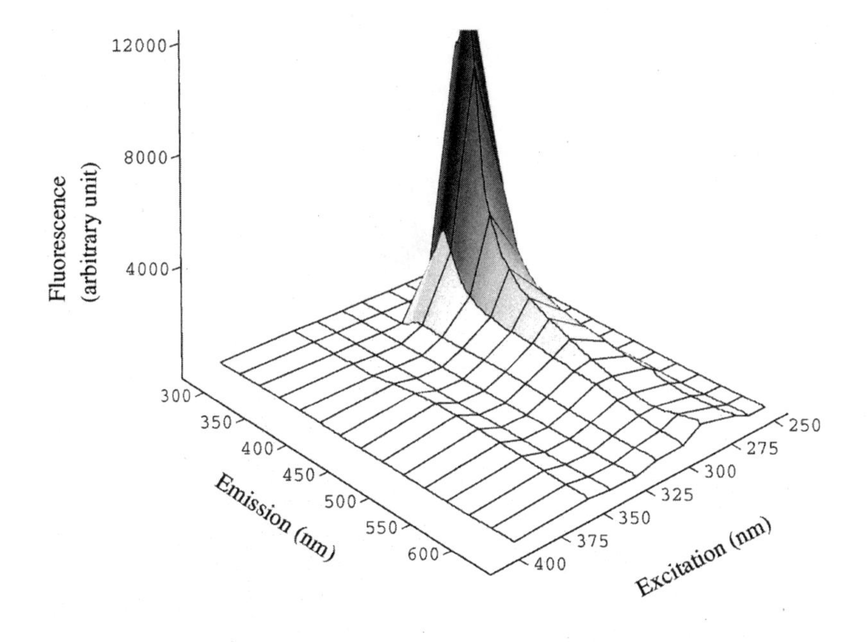

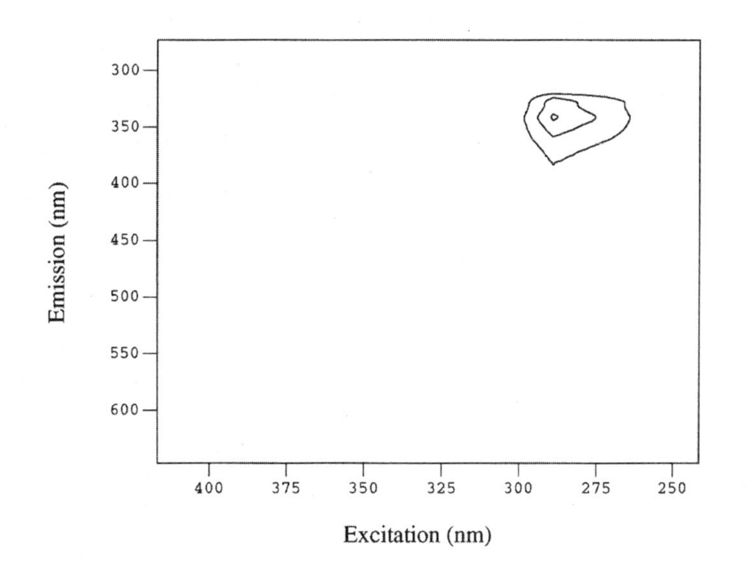

OANGB CY4J 234 cm

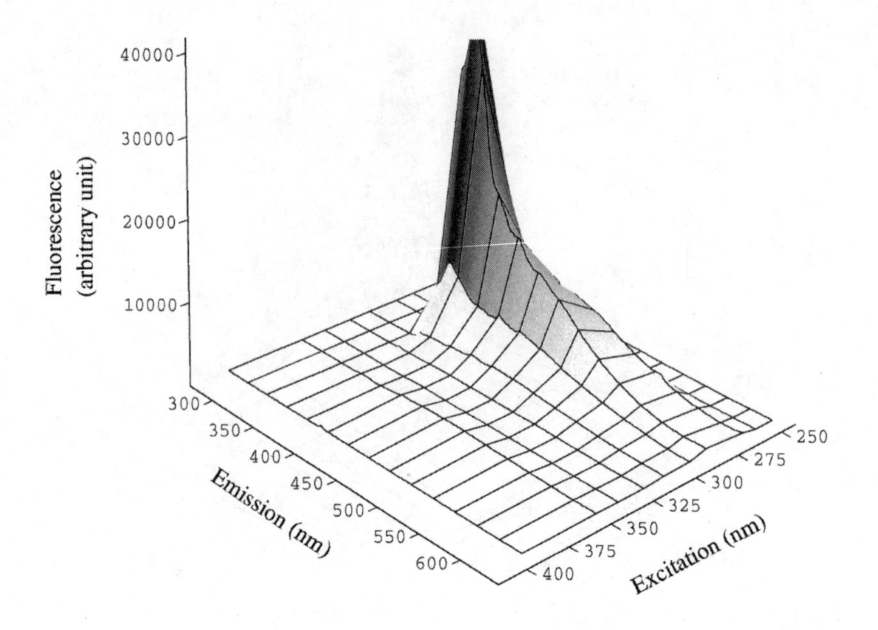

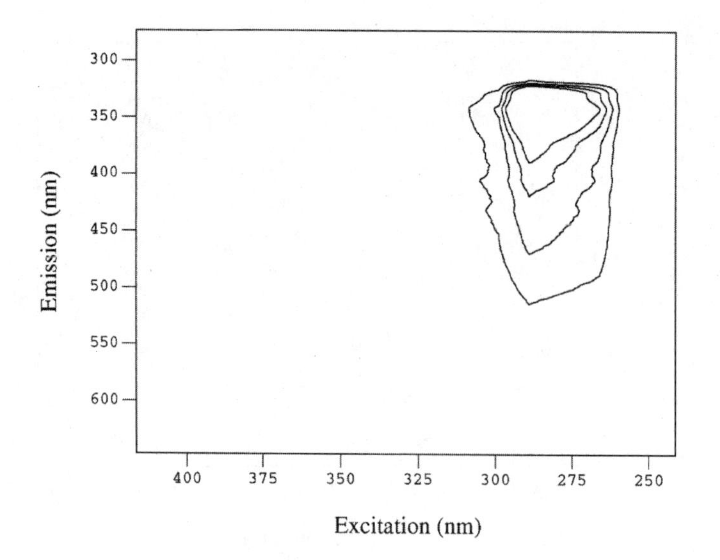

OANGB CY4J 314 cm

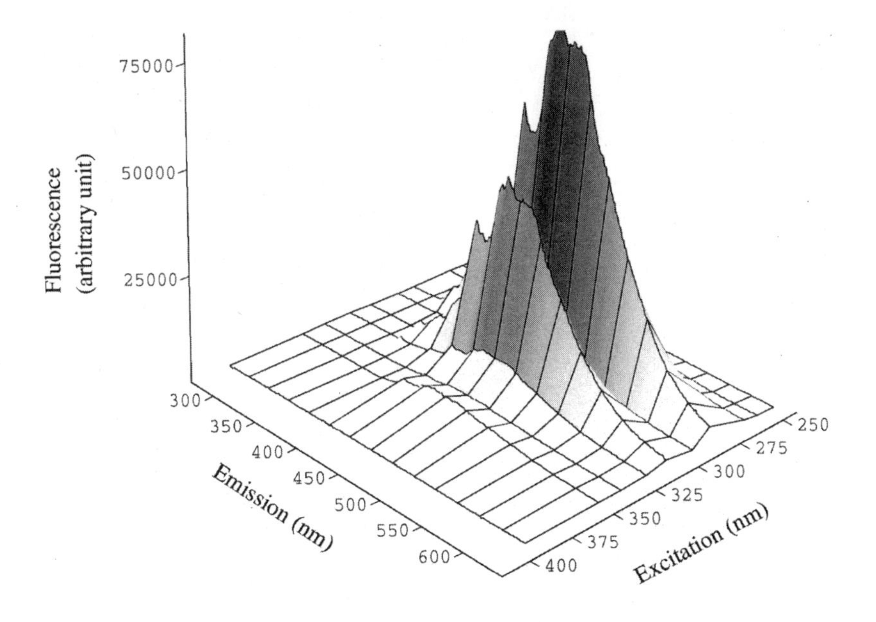

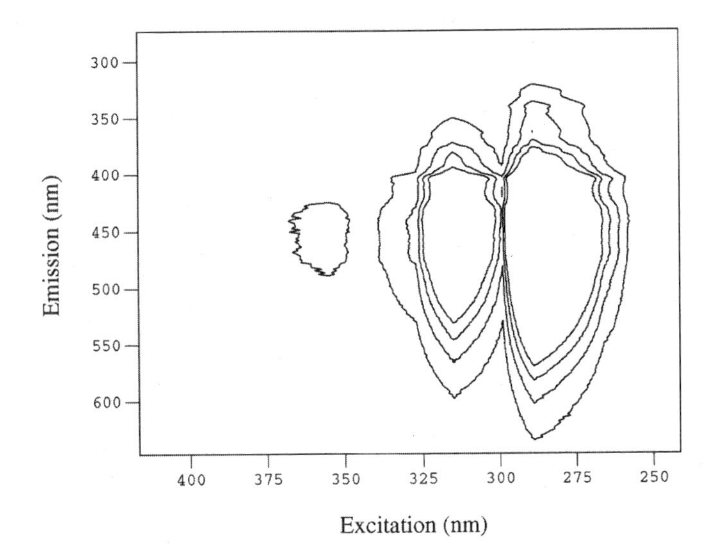

OANGB CY4J 652 cm

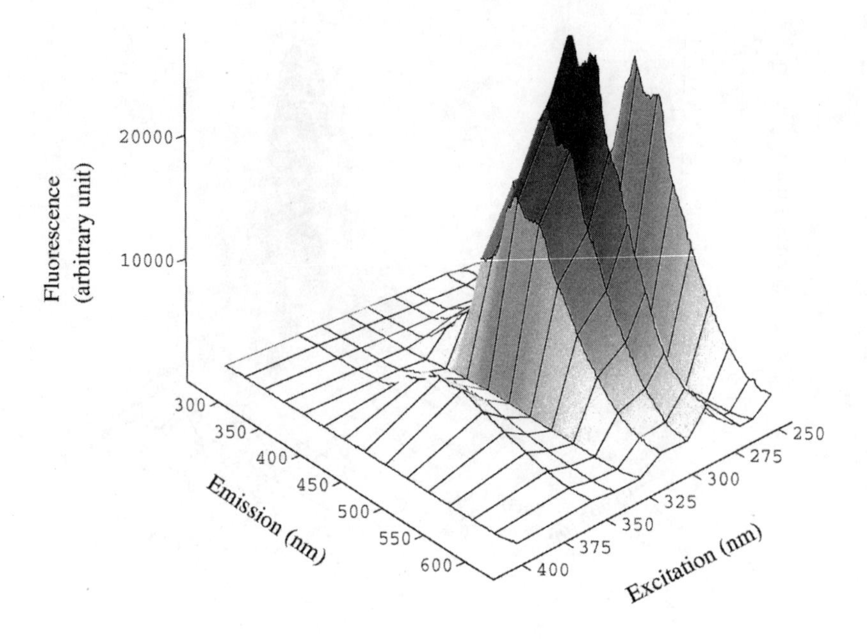

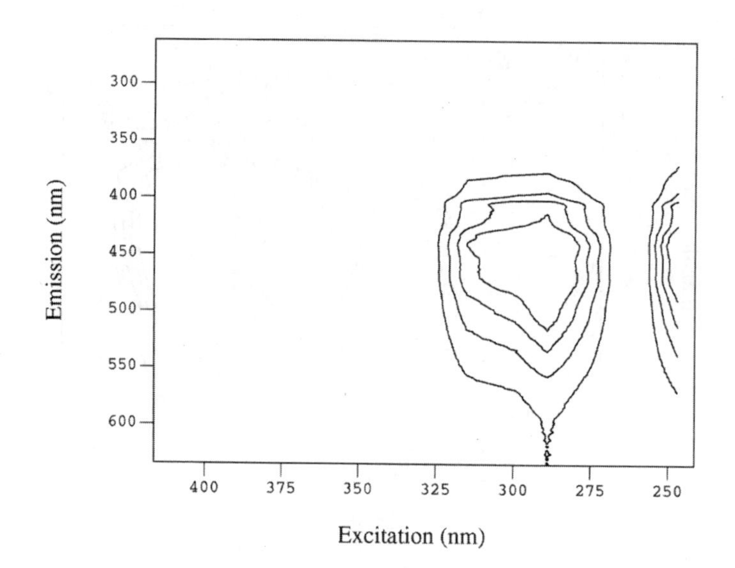

OANGB CY4K 174 cm

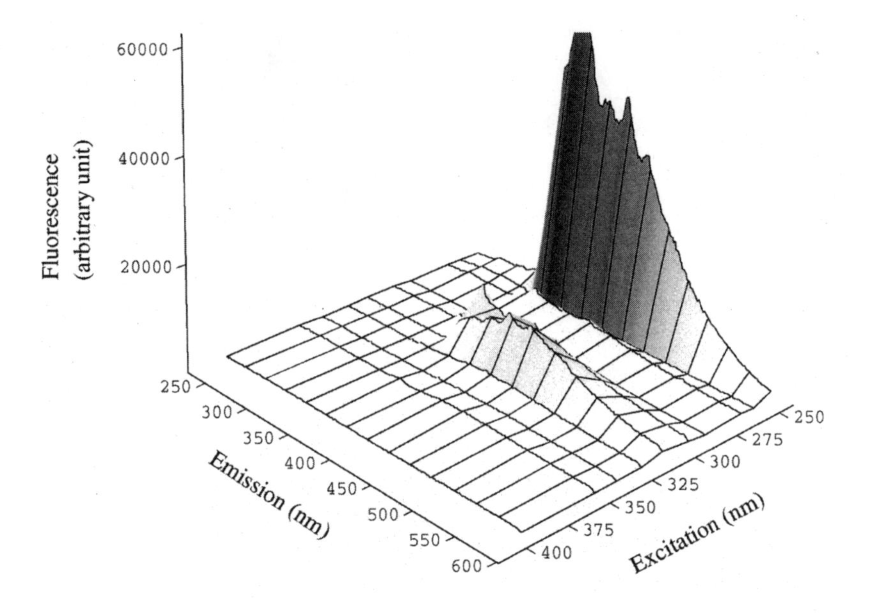

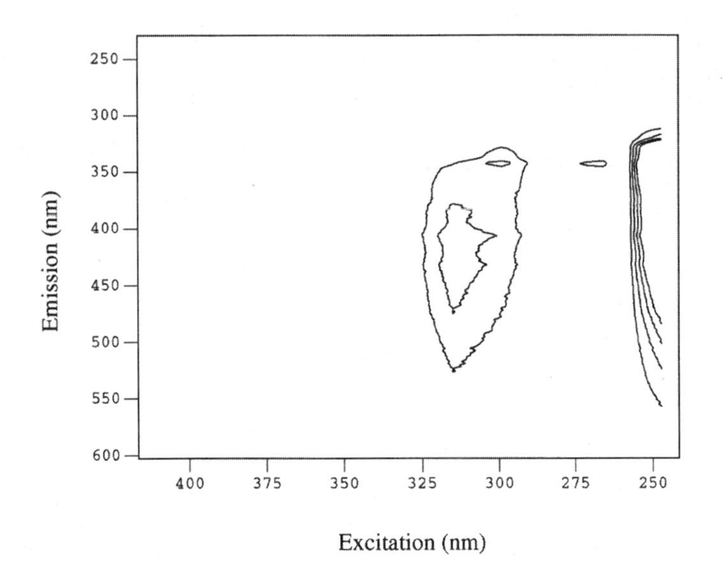

OANGB CY4M 190 cm

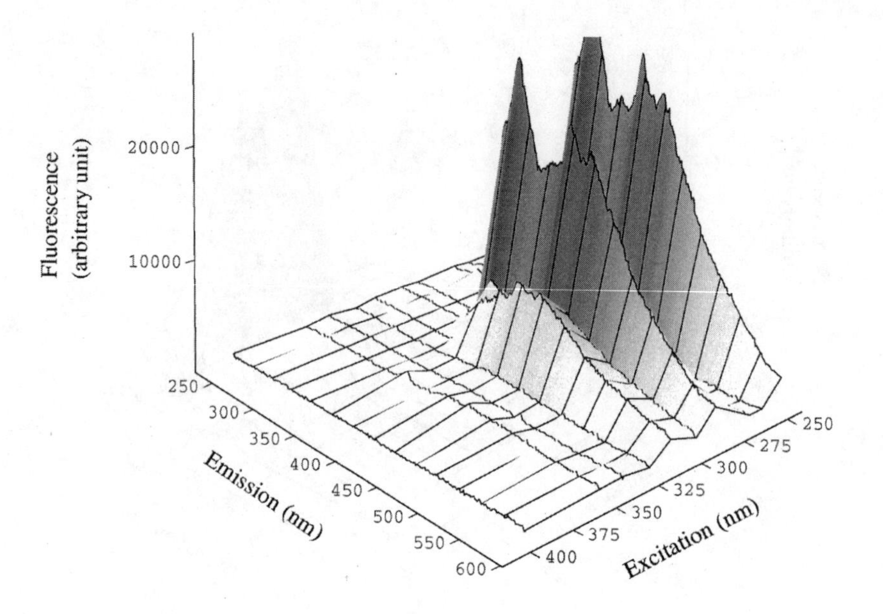

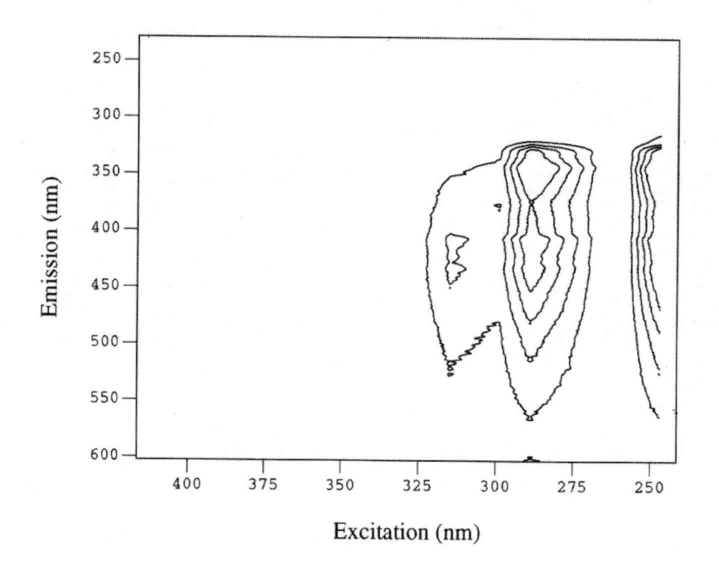

OANGB CY4P 148 cm

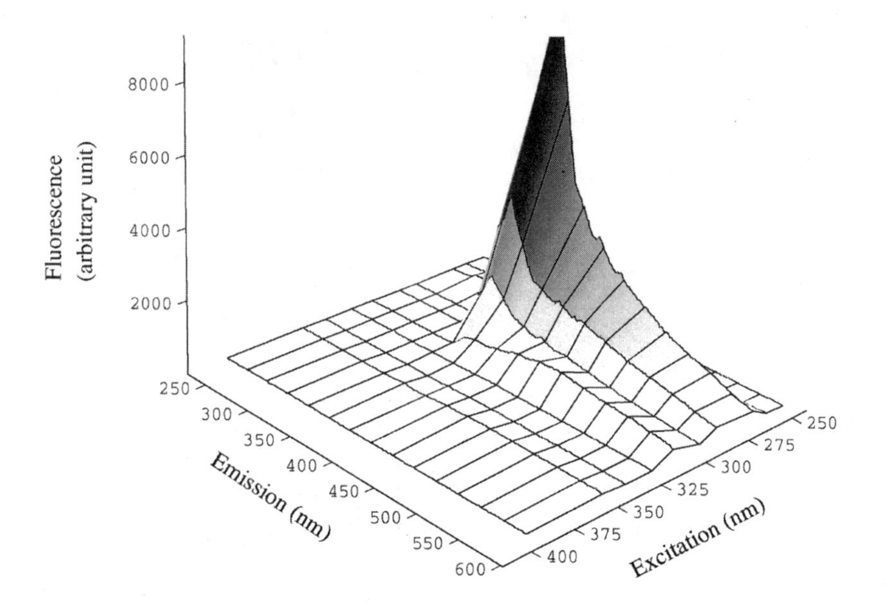

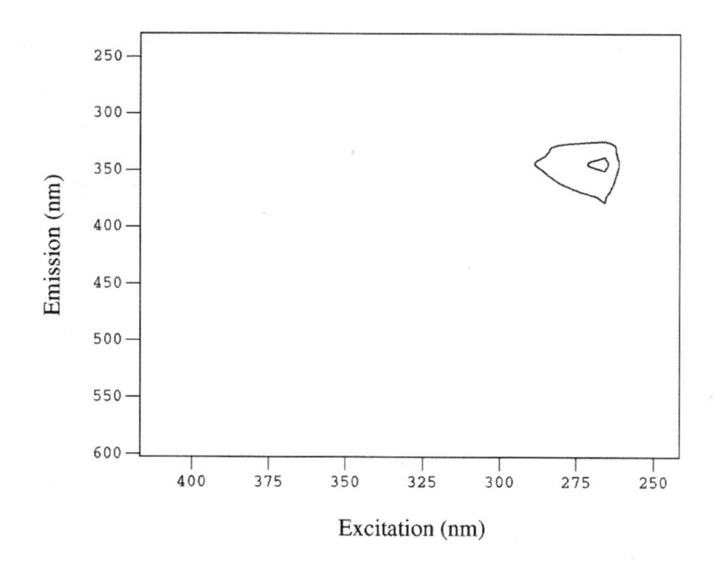

OANGB CY4P 258 cm

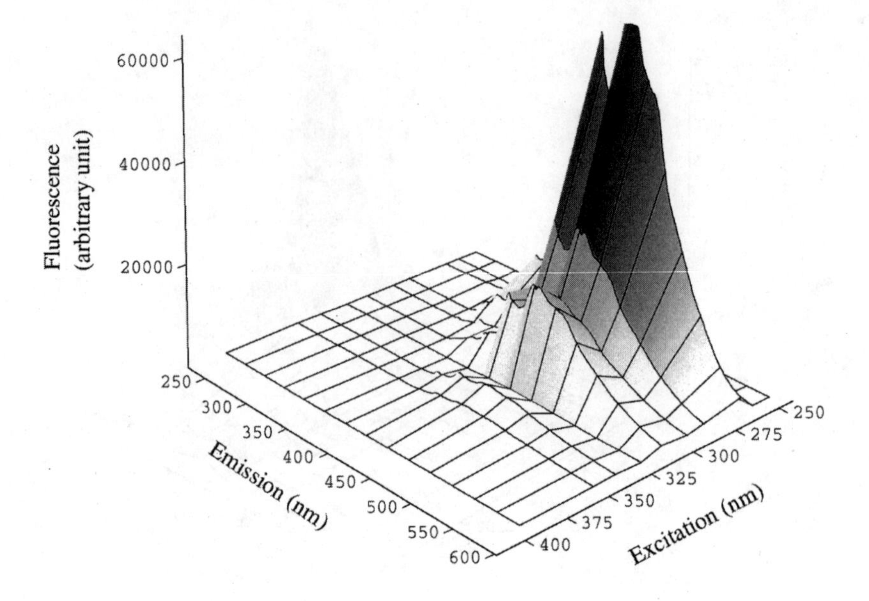

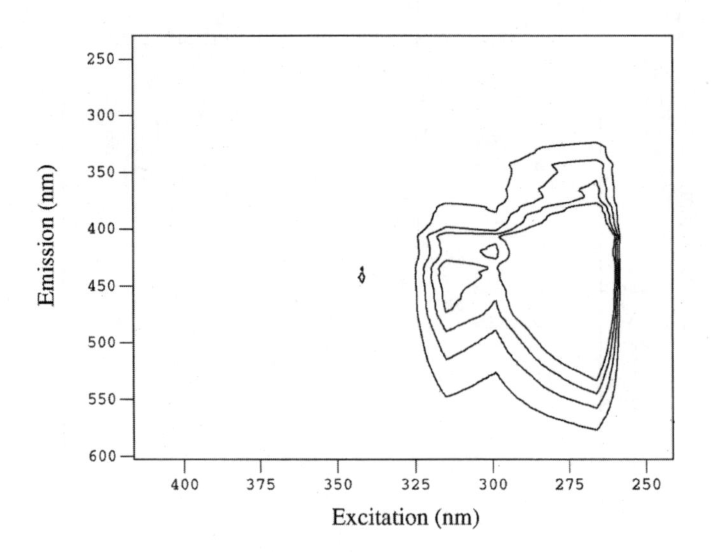

OANGB CY4P 714 cm

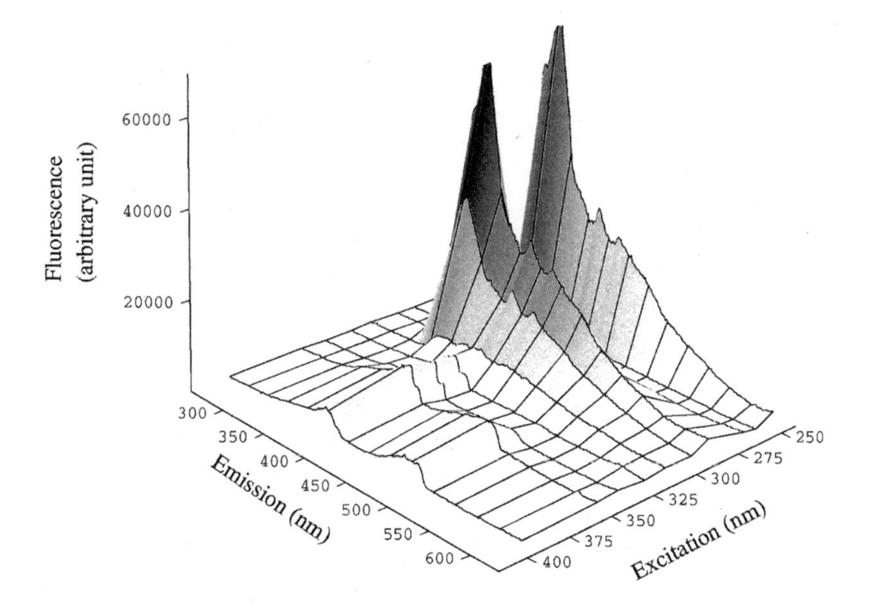

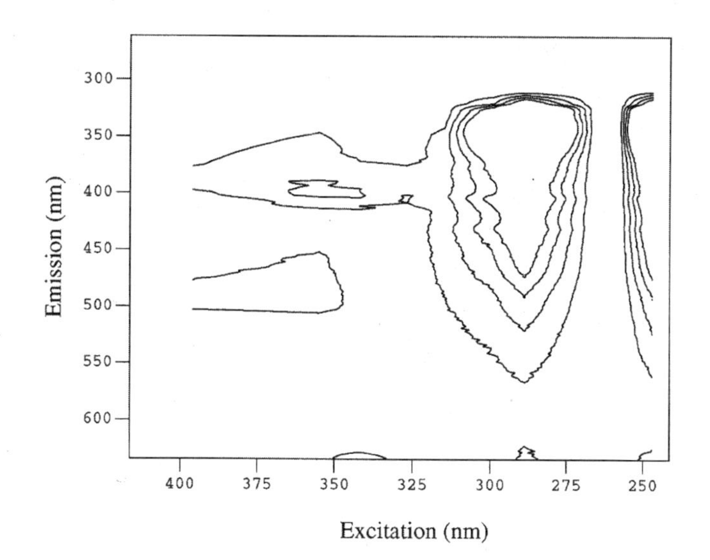

OANGB CY4Q 214 cm

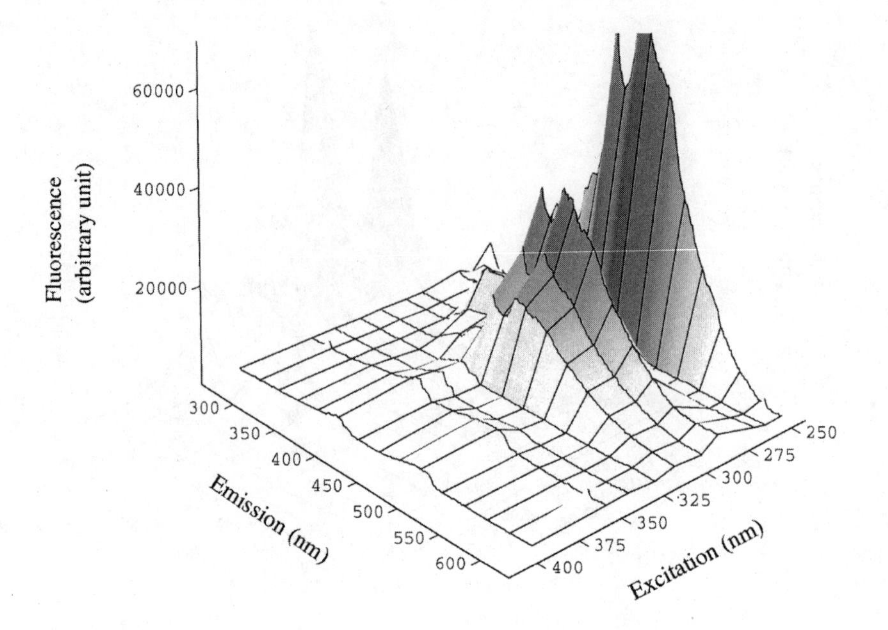

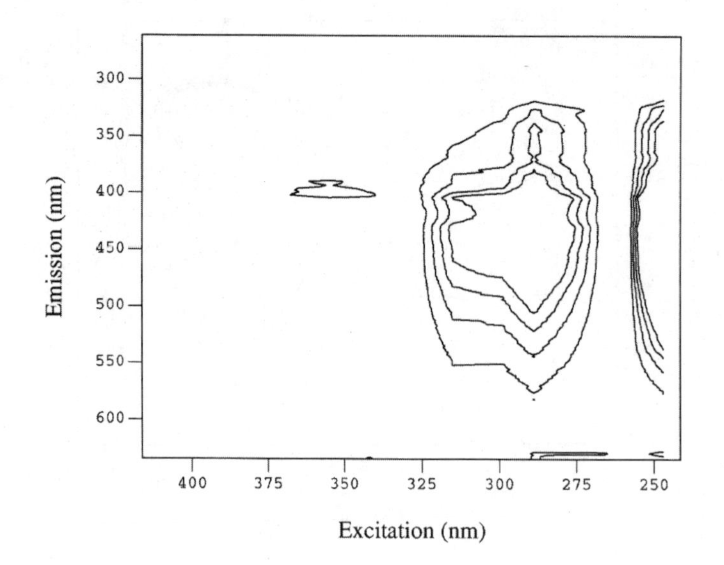

OANGB CY4Q 672 cm

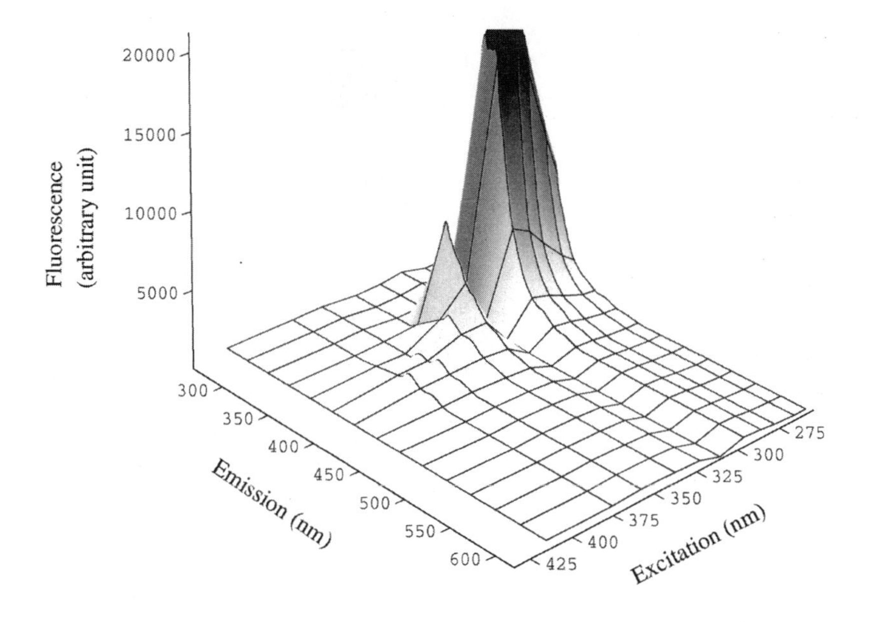

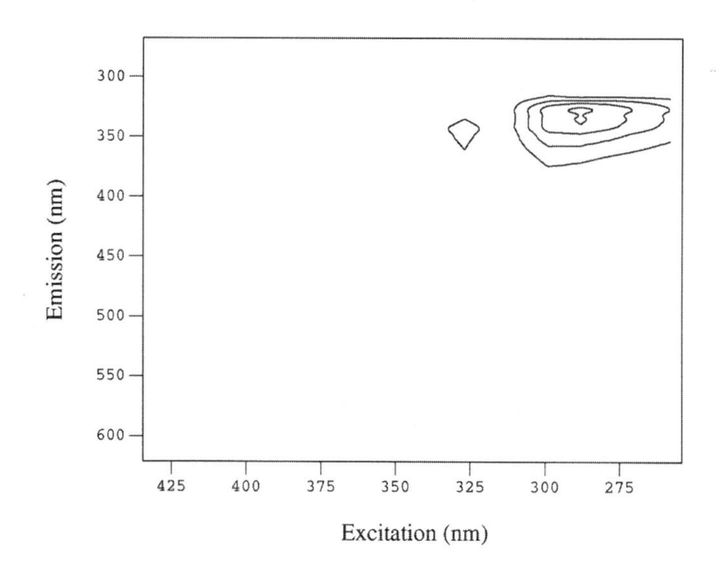

Gasoline (neat)

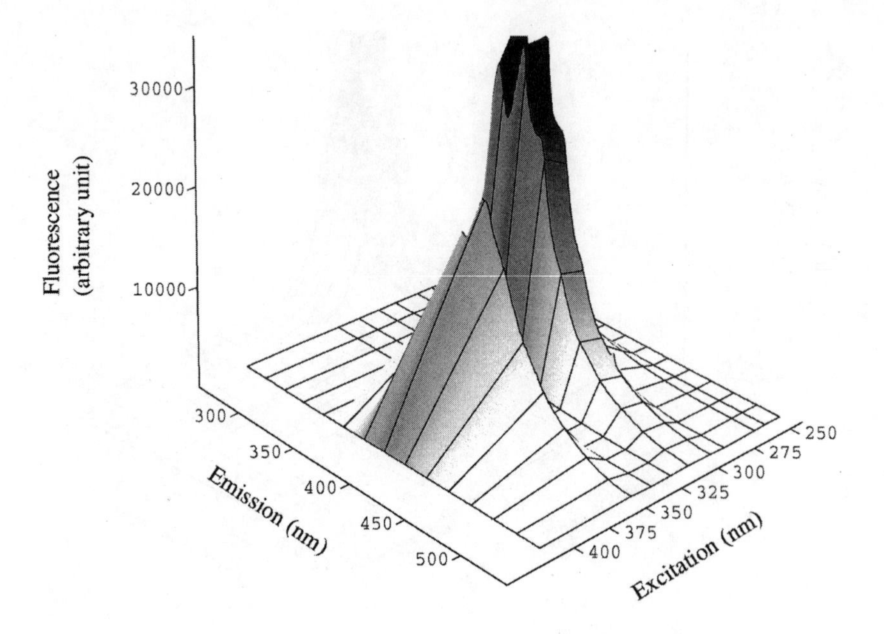

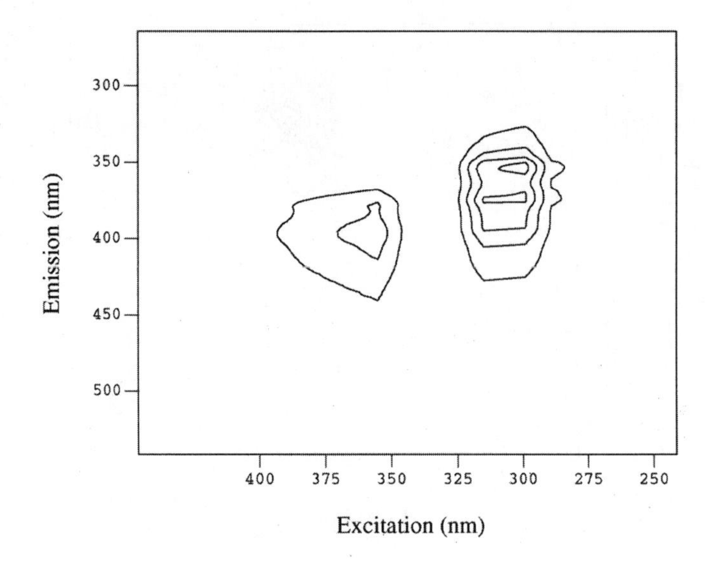

Heating oil (neat)

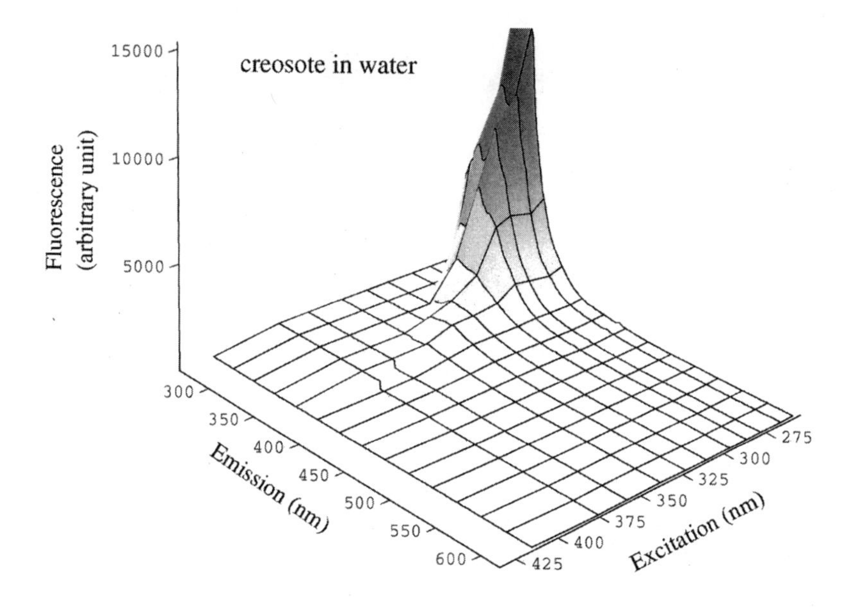

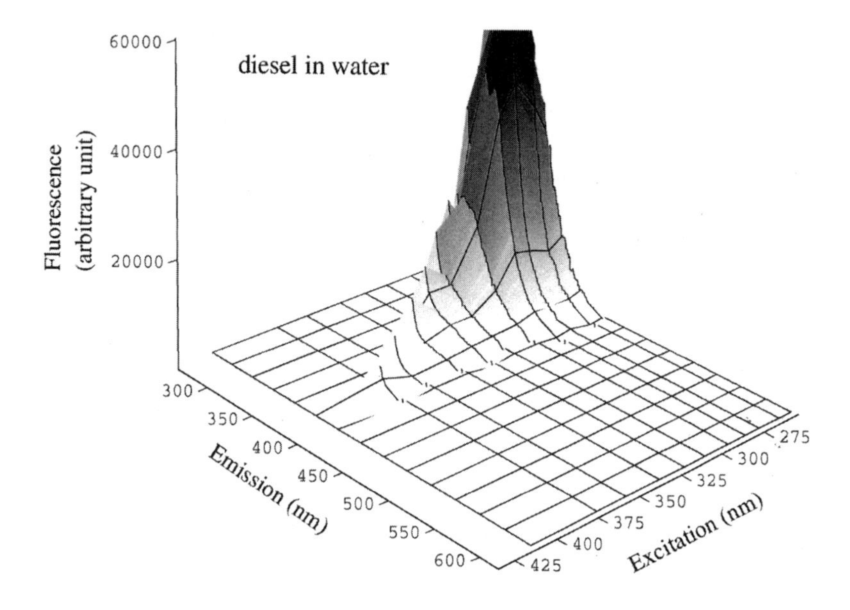

Creosote in water and diesel in water

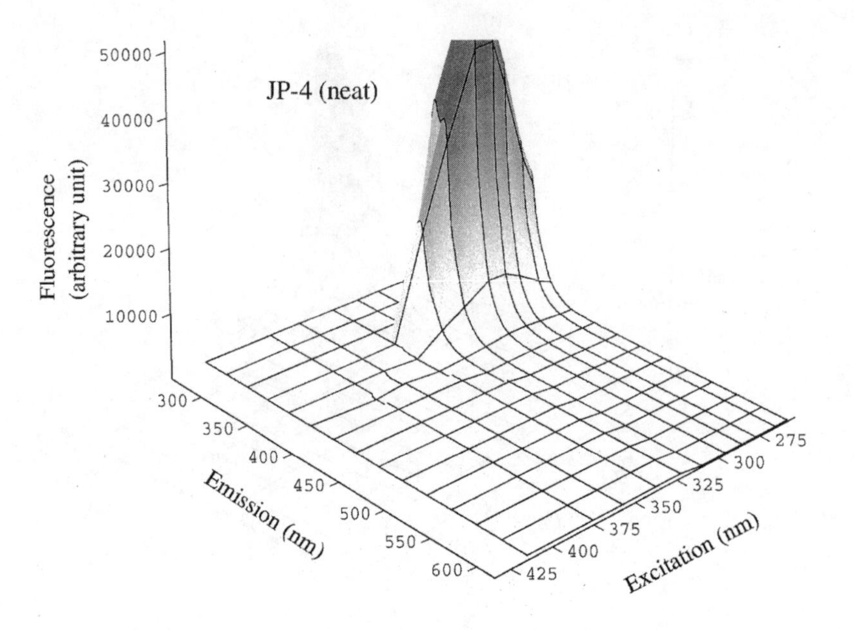
JP-4 (neat)
Fluorescence (arbitrary unit)
50000
40000
30000
20000
10000
300
350
400
450
500
550
600
Emission (nm)
275
300
325
350
375
400
425
Excitation (nm)

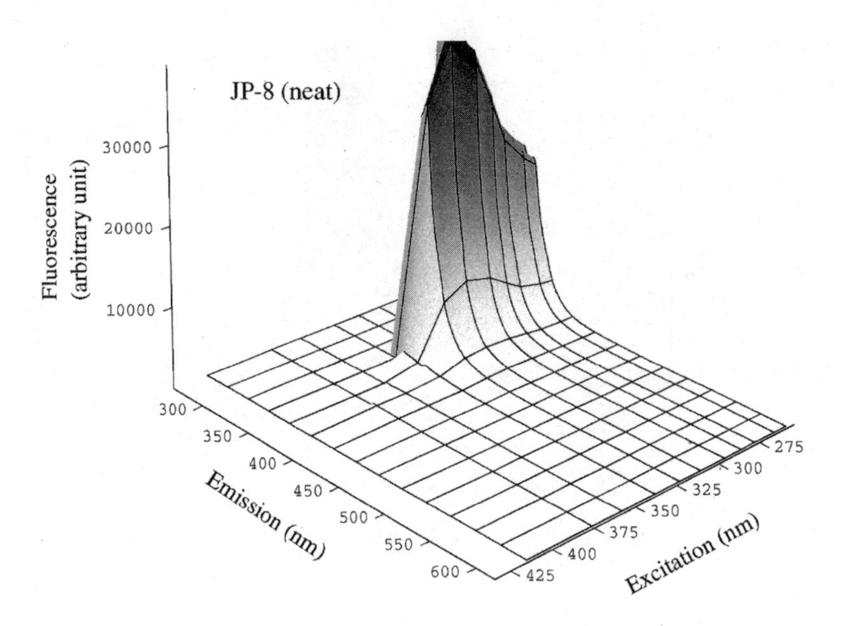
JP-8 (neat)
Fluorescence (arbitrary unit)
30000
20000
10000
300
350
400
450
500
550
600
Emission (nm)
275
300
325
350
375
400
425
Excitation (nm)

JP-4 and JP-8

Index

E

F

G

H

I

J

K

L

M

N

O

P